High Sulfur Coal Exports
An International Analysis

Edited by
Michael M. Crow

With a Foreword by Senator Charles H. Percy

Published for
Coal Extraction and Utilization Research Center

Southern Illinois University Press
Carbondale and Edwardsville

Technical Information Coordinator: Rhonda Vinson
Technical Editor: Herbert Russell
Word Processing Supervisor: Linda Harris
Word Processing Operator: Rick Holt
Information Specialist: Donna Davin
Production Supervisor: John DeBacher

First Printing July 1983

Library of Congress Cataloging in Publication Data

Main entry under title:

High sulfur coal exports.

   Proceedings of a conference held June 8-9, 1981 in Carbondale,
Ill., and sponsored by the United States Senate, Committee on
Governmental Affairs, Subcommittee on Energy, Nuclear Proliferation
and Government Processes.
   Includes bibliographical references.
   1. Coal trade--Congresses. 2. Coal trade--United States--
Congresses. 3. Coal--United States--Sulphur content--Congresses.
4. Sulphur dioxide--Environmental aspects--Congresses. I. Crow,
Michael M. II. United States. Senate. Committee on Governmental
Affairs. Subcommittee on Energy, Nuclear Proliferation, and Govern-
ment Processes.
HD9540.5.H53 1983   382'.4224'0973   83-4657
ISBN 0-8093-1122-4

CONTENTS

84-5491

FOREWORD

It has been said that the world has completed an era in its energy history. The energy situation we face today is a grave one. No longer satisfied with the rising prices and the vulnerability of imported oil from the Middle East, nations throughout the world are now aggressively pursuing new energy strategies to carry themselves through the 1980s and beyond. The free world is cutting back on oil imports and relying more on energy available from more stable sources of supply. One of these stable sources, both for our country and for the world, is America's own vast domestic reserves of coal.

The United States possesses several hundred billion tons of demonstrated coal reserves, nearly one-third of the world's total. The State of Illinois alone accounts for 68 billion tons, representing the third largest reserves in the nation. This abundance of coal means that Illinois and our entire nation, and all our people, can expect to profit from this worldwide shift toward coal.

Yet, in 1980, foreign coal colliers which contracted to receive American steam coal were forced to wait as long as 100 days for delivery. Storage terminals, built to accommodate smaller shipments of metallurgical coal, were unable to accept steam coal deliveries in the amount required. Antiquated rail links and shallow ports were equally ill-prepared for the challenge. Real doubt began to emerge among foreign buyers about whether the United States, despite our ample coal reserves, could be trusted as a reliable energy supplier. Ironically, the Soviet Union, with its planned natural gas pipeline to Western Europe, is now seen by some Western nations as a more reliable supplier of energy than the United States. If a good domestic infrastructure is established, America would demonstrate itself as a reliable supplier, and our coal could satisfy a large portion of the energy needs of many nations. Estimates suggest that the United States will be capable of exporting as much as 386 million tons of coal per year by 2000.

As chairman of the Foreign Relations Committee and the Subcommittee on Energy, Nuclear Proliferation and Government Processes in the United States Senate, I am doing all I can to promote coal for export to markets thoughout the world. Illinois, with its enormous reserve base and relative proximity of transportation, water, and fine labor force, is a state strategically important to the future energy needs of nations. On June 8th and 9th, the Governor of Illinois and other top state officals, executive officers from

American coal companies, labor officials, and dignitaries from Europe and the Far East joined me in Carbondale, Illinois, to discuss this important global issue. Such topics as developing international markets and marketing mechanisms for America's coal, evaluating the domestic potential for American coal production and exports, and investigating Illinois coal's potential as an emergency fuel in the event of an international energy shortfall were addressed and discussed during the U.S. Senate Subcommittee field hearing and conference.

The consensus of the High Sulfur Coal Export Conference and Hearing was clear: unless we send a strong signal <u>now</u> to foreign coal buyers--telling them we have goods to sell and that we can be counted on to honor long-term contracts--we are going to lose international business to other energy exporting nations that <u>do</u> have the resolution to act. Let's tell the world that Illinois coal is immediately accessible to the Mississippi River, has a massive network of rail links, and can be shipped to their countries with <u>far</u> less waiting time than those along the East or West Coast.

Certainly, Illinois Basin coal cannot meet the needs of all coal importers. Its sulfur content places some limitations on its uses. But these restraints should be seen as extra marketing challenges, <u>not</u> as barriers to success. Illinois Basin coal producers may have to work harder to develop markets for their product, but those markets can be found. It will take some time, of course, but we have reason to be optimistic about what we can all accomplish by working together.

Charles H. Percy
United States Senator

PREFACE

The problems surrounding the use of America's high sulfur coal resources are staggering given the international goal to maintain a clean environment and our goal to move away from unstable expensive oil. High sulfur coal by its very nature produces compounds during utilization or conversion which are deleterious to human health. This fact is known. It is also known that because this supply of easily transportable liquid and gaseous fuels is finite and because the world's demand for electricity is growing at a rapid rate the demand for fuel on a world scale has increased dramatically in recent years. Compounding this situation is the fact that the supply of liquid fuels in particular is not secure. This lack of a stable secure fuel has led many industrial nations and those wishing to continue industrialization to search for a more secure and stable fuel source. It is the contention of many analysts that American coal can meet the requirements for selection--including price, stability, and deliverability specifications.

There are, however, problems. Much of the higher quality U.S. coals also contain a high sulfur content. Fortunately, these coals are located in a region where several transportation options are available and could, with some infrastructural and technical changes, become a leading fuel for the world energy market.

In an effort to define the major barriers to developing high sulfur U.S. coals as a world fuel, leaders and scientists from around the world met for two days (June 8-9, 1981) of intensive dialogue on the subject. Representatives from U.S. coal producing firms, transportation firms, foreign buyers, international energy groups and various universities and governments met to confront the issue at hand and to determine how U.S. high sulfur coal could be brought into the world energy market as a low cost, environmentally acceptable alternative to high cost, unstable OPEC oil.

A conference and an associated field hearing of the United States Senate Subcommittee on Energy (Committee on Governmental Affairs) provided the forum for interaction between major producers, policy makers, and potential consumers of U.S. high sulfur coal. During the two-day event, Asian and European government leaders met and openly discussed their problems with purchasing U.S. high sulfur coal. In response, U.S. coal mining, coal processing, transportation, financial and management experts defined the problems and identified areas for potential interaction.

Since the June 1981 event much has happened to forward the future development of high sulfur coal into the world energy market. In the State of Illinois, an Office of Coal Commerce was established to coordinate coal export activities. Between the U.S. government and several European and Asian governments negotiations continue regarding the acquisition of U.S. coals as a primary fuel for their national energy economies.

It is obvious that an international conference held at Southern Illinois University at Carbondale could not have been conducted without the help of many individuals and groups, beginning with the distinguished presenters, panelists and participants. Further, the work of several key individuals helped significantly to make this event and this document a reality. U.S. Senator Charles Percy and Congressman Paul Simon conducted the hearing and moderated panels during the conference. Invaluable moral and material support was given by Dr. Lyle Sendlein, who was Director of the Coal Extraction and Utilization Research Center at SIUC. Working in Washington, coordinating the Senate hearing and contributing extensively to the concept and general direction of the event was Joshua Levin of Senator Percy's staff. In Carbondale, Charles Pizer and Timothy Crawford coordinated international relations and conference operations. In addition, funding was provided by the Illinois Institute of Natural Resources, now the Illinois Department of Energy and Natural Resources, the Illinois Department of Commerce and Community Affairs, and the President's Office of Southern Illinois University at Carbondale.

In the preparation of the proceedings of the High Sulfur Conference, Dr. Herbert Russell, Technical Editor at SIUC Coal Research Center, spent innumerable hours assisted by Floyd Olive editing the testimony and papers. In addition, the Coal Research Center's Office of Technical Information, coordinated by Rhonda Vinson and staffed by Linda Harris, Debbi Crain, Donna Davin, Herbert Russell, and Rick Holt, actually prepared the conference documents.

The material herein is not representative of the total activity. The entire record for both the U.S. Senate Hearing and conference is available from the Senate Subcommittee on Energy of the Governmental Affairs Committee.

Michael M. Crow
March 1983

Utilization of Coal in Energy Transition and in Crisis Situations

BACKGROUND AND OVERVIEW
ULF LANTZKE

Coal offers one of the most important opportunities to make the transition from an oil-based economy successful and it has been a central part of the International Energy Agency (IEA) program over the last several years. In 1978, the IEA published Steam Coal Prospects to 2000, which concluded that coal use and production should be encouraged to more than double by the year 2000. A new world steam coal trade was viewed as essential in achieving this objective.

To spur coal development and to build confidence between the major producing and consuming countries, IEA Energy Ministers agreed in May, 1979, to Principles for IEA Action on Coal. The principles created a mutually accepted political framework to promote the rapid expansion of coal and a world steam coal trading system.

To provide for more government-business cooperation at the international level, the IEA established a Coal Industry Advisory Board (CIAB) composed of senior executive officers of companies with an interest in coal either as producers, consumers, or builders of the coal infrastructure. The Board gave its first report to IEA Ministers in December, 1980.

These activities gave IEA an international perspective on both the problems and the potential for expanding production, as well as the use and trade of coal, and I would like to review some of our major conclusions.

LONG-TERM OUTLOOK AND THE ROLE OF COAL

The 1980s and the 1990s must be a period of major transition toward energy economies no longer dominated by oil use nor dominated by any other single resource, but being characterized by a much more balanced structure. With strong policies to promote coal, nuclear energy, and conservation, the IEA estimates that oil as a percentage of IEA countries' total energy supply can be reduced from 50% today to about 25 to 30% by the year 2000. This means that oil imports into IEA countries can be reduced from 22 million barrels of oil a day (mbd) today to 15 mbd by the year 2000. To achieve this scenario, IEA coal production and use would almost have to double by 1990 and triple by 2000. World trade in steam coal would have to increase tenfold--to about 700 million tons of coal.

Coal could meet almost two-thirds of the additional energy requirements in IEA countries between now and the year 2000. Such an increase in terms of energy would be comparable to present OPEC production. World trade alone in coal would be about the energy equivalent of present day Saudi Arabian oil exports. Coal could surpass oil as a contributor to IEA energy supplies, increasing its share from 20% today to 35% in the year 2000.

Achieving this type of future will not be easy and it still is uncertain if it can be achieved. The IEA Coal Industry Advisory Board has concluded that a doubling of coal use by 1990 and a tripling by 2000 will not happen unless the production and transport infrastructure is expanded in a timely manner and more activity takes place on the user side to substitute coal for oil in electric generation and industry. Nevertheless, there has been considerable activity over the last two years. Coal demand is high; coal production, despite some recent problems in the U.S., Poland, and the United Kingdom, is also swinging upward. This suggests to me that we are witnessing a coal revival and this early momentum must be sustained.

THE ROLE OF THE U.S. AND COAL EXPORTS

The only way industrial countries can reach this goal is with a major U.S. commitment to export steam coal. U.S. annual coal production of 1.2 billion tons and exports of 150 million tons by 1990 are urgently needed. By the year 2000, we look for U.S. production to be about 2 billion tons and exports to increase to 300 million tons yearly. This would require a tenfold increase in surface mine production in the west, as well as at least a doubling of eastern underground coal mining.

With regard to U.S. ports, an infrastructure would have to be built to transport coal to potential domestic as well as overseas markets. In 1980, U.S. exports of coal were up by 40%. The American coal industry responded, but congested port facilities resulted in high costs that were borne by the buyers as ships waited to be loaded. A major expansion of loading facilities to handle large coal-carrying ships is essential.

The United States should work with potential coal-using countries in Europe and Japan to build confidence that the U.S. is interested in becoming a reliable coal supplier. Prompt attention to port and other infrastructure problems would assure U.S. trading partners of our intentions.

For the U.S., economic, political, and strategic benefits could be substantial for the following reasons:

1. coal can provide the U.S. and other Western economies with an alternative to continued oil imports from uncertain sources;
2. coal exports would improve the U.S. balance of payments; and
3. for local communities, coal would mean new jobs and increased tax revenues.

HIGH SULFUR COAL

Given this general world and national outlook, I am pleased to see so much interest in this part of the country to expand both your domestic and overseas market. High sulfur coal is rich in energy content, and studies made by the IEA Coal Research Center in London suggest that sulfur oxide emissions can be kept to environmentally acceptable levels with known technology.

However, it must be recognized that the problem of protecting the environment is a worldwide concern. Generally, it has developed objectively as opposed to being solely an emotional or political issue. It has become an element of value judgement in making the difficult choices and trade-offs between energy development and maintenance of viable environmental conditions of life. The Clean Air Act in the U.S. is one reflection of this concern. Similar types of views exist in export markets.

This problem is severe and can only be overcome by new technologies. As a consequence, further demonstration of known technological paths must be developed, not only in the U.S., but also in potential export markets if high sulfur coal is really to have the role which may be needed for it to provide a contribution to more balanced world energy markets.

EMERGENCY UTILIZATION OF COAL

I would like also to say a few words about the emergency utilization of coal in trying to answer the basic question to what extent coal could be substituted for oil in a short-term crisis. Our knowledge in this area is limited. Normally, at the IEA we have viewed coal as primarily a long-term alternative. Lead times are long in getting the world steam coal trade infrastructure built, and until now I do not believe that anyone had imagined a situation where U.S. coal, for example, might be supplied to Europe or Japan in a short-term crisis. I believe, however, that this question deserves attention. For an oil crisis of only a few months, the response may take too long to substitute coal internationally, although every effort should be made in those countries, like the U.S., which have coal resources at home, to substitute those resources domestically. On the other hand, if we do get into a situation of chronic oil shortages--which might occur during the mid- to late-1980s--then moving adequate quantities of coal in a timely manner may be essential to sustain the economies of the West.

IEA CONCLUSIONS ON COAL

First, coal, which once fueled the Industrial Revolution, is gathering momentum to provide the single most important energy contribution to the future of modern societies. Given uncertainties about the future role of nuclear power and the long lead times for commercialization of new and renewable energy sources, the conclusion is inescapable that coal must play a vital role in our energy future.

Second, the challenge of the next 20 years for the increased international utilization of coal is to mine, transport, and trade enough coal to meet IEA coal demands equivalent to 40 million barrels of oil per day. This is a large increase compared to the present level of IEA coal requirements which are equal to about 15 million barrels of oil per day. The potential exists, but there are serious obstacles to be overcome.

Third, political support at the highest level is already behind the coal option. But general commitments need to be supported by specific measures of consistent, long-term coal policies. For success, close cooperation between industry and government will be essential because, while governments provide the framework in which national economies and new energy systems develop, the responsibility for getting the job done rests with industry. The economics are favorable; the technology is available; and while major environmental problems in production and use exist, they are, in my view, solvable.

VULNERABILITY OF OIL CONSUMERS
PETER BORRE

Oil has been the crucial energy source for the industrial countries. However, the future of international oil supply is uncertain. For a number of political and economic reasons, future OPEC oil production is unlikely ever again to exceed the 1980 high point of 29 MMB/D. It could drop below current production of 26 MMB/D. But even if OPEC production were not to decline, exports would be reduced as OPEC countries will likely increase their domestic consumption. Although production will increase in some non-OPEC developing countries, such as Mexico and Egypt, this production is likely to be offset by an increased demand for oil in those and other developing countries.

The implications of this outlook for the world price of oil over the next decade will depend, of course, upon a few key factors including the following: the rate of economic growth; the continuation of the gains in energy efficiency and conservation stimulated by the sharp increases of the price of oil; the pace of development of alternate sources of energy supply; and the future prospects for supply disruptions. The real price of oil conceivably could remain stable, or it could continue its upward trend, as economies recover and additional conservation becomes more costly to achieve. If supply disruptions occur, the price of oil could increase sharply.

Oil importing countries will remain highly vulnerable to oil supply disruptions. We have already experienced a number of oil disruptions since 1973 including the following:

1. the 1973-74 Arab oil embargo of the United States and the Netherlands which reduced world supplies by about 400 million barrels during the period from November 1973 to March 1974;

2. the 1979 revolution in Iran which reduced supplies in the world oil market some 250 million barrels in the period from January to March 1979 and helped increase oil prices by over 100% in one year (from $13-$14/barrel on January 1, 1979, to a level of $28/barrel on January 1, 1980); and

3. the 1980 Iran/Iraq war which reduced oil supplies available to the world by about 200 million barrels in the fourth quarter of last year.

The costs to the world economy, in terms of increased unemployment, additional inflation, and lowered economic growth were substantial.

COSTS OF OIL SUPPLY DISRUPTION

There were also significant international political costs such as the increased leverage of oil exporting countries. For these reasons, the Administration has undertaken to deal with the potential jeopardy to our energy security, our national security, and our foreign policy interests through development of contingency response plans aimed at improving the ability of the United States and our allies and friends to deal with significant disruptions, and also aimed at reducing our vulnerability in the future. Our policies will rely on market forces to clear disruptions, rather than utilizing restrictions such as allocation and price controls, which have proven counterproductive in the past. We will also work closely with our allies and friends to assure that collective actions by oil consumers will reinforce, rather than work against, one another.

U.S. ENERGY PLANS

Although reducing U.S. vulnerability to supply disruptions is a first priority, we recognize that, in the longer term, lowering the dependence of the U.S. and its allies on oil imports, particularly from insecure areas of the world, is also important to our national security and economic well-being. Therefore, this Administration is also taking steps to increase our domestic energy production and to improve our energy efficiency. Again, we will rely on the market rather than government programs. Government policy will be to provide the appropriate market environment through the decontrol of oil and the lowering of regulatory barriers to the increased production, transportation, and use of domestic energy sources. The outlook is positive. One reason is our extensive coal resources; another is the creative and competitive nature of the U.S. coal industry.

Coal is our most abundant energy resource. It is estimated that we have a resource base of nearly 4 trillion tons of coal, of which 475 billion tons are economically recoverable, using existing technology; this reserve base is amply sufficient to supply America's existing needs for hundreds of years.

COAL EXPORTS AND ENERGY SECURITY

Even if we could reduce substantially our reliance upon imported oil, we must be concerned about the energy security of our allies and friends overseas, including their vulnerability to disruptions and their dependence upon unreliable energy supplies. The United States can make two contributions to assist our allies' energy security:

1. by reducing oil imports, the United States as the world's largest oil importer makes a positive contribution to reducing price pressures on the world oil market; and
2. by expanding exports of steam coal, the United States contributes to our allies' efforts to reduce their continuing high levels of oil import dependency and vulnerability.

RECENT WORLD DEMAND FOR STEAM COAL

In contrast to oil, which may be supply-constrained, the world's coal reserves and resource base have the potential to satisfy a large and rapidly growing demand. Coal probably has more potential than any other resource for providing additional energy, until the transition to renewable energy evolves. Every indicator points to a significant increase in domestic and international coal demand, both in the short and long term. In fact, the surge in U.S. exports that was forecast by the 1980 World Coal Study to begin in 1985 has already begun. The World Coal Study projected that by 1985 the United States would be exporting 20 to 30 million tons of steam coal. We reached this level last year by exporting nearly 27 million tons of steam coal, 16 million tons to countries outside North America, and about 11 million tons to Canada. Exports of metallurgical coal in 1980 amounted to 63 million tons. Total steam coal exports unexpectedly jumped 90% over the 1979 level; exports to Europe alone increased by 700%. This surge was caused by several factors, among them the following:

1. Polish coal production declined and exports fell by 25%;
2. an Australian strike disrupted that country's exports; and
3. countries and companies overseas moved to increase their supplies of non-oil energy as new electric power and industrial coal-burning facilities became operational.

While Australian exports have recovered, the prospects for building up substantially Polish coal exports may not be particularly promising, at least in the short term.

We all are aware of the impact that the surge in demand in 1980 had on U.S. coal export facilities, causing delays and increasing purchasers' costs. We should not let these problems, which can be corrected through the interplay of market forces, obscure the long-term prospects.

U.S. SHARE OF THE WORLD STEAM COAL MARKET

As the world demand for coal increases, United States exporters will participate significantly in world steam coal trade for the remainder of the century and beyond; both Europe and Asia will be reliable customers.

Europe presents the greatest opportunity for U.S. coal exports for near- and mid-term sales. It is the largest and most diversified market, with significant growth expected in coal-fired electric power facilities and industrial plants. There has been a recent growth of demand for higher sulfur coal in the cement industry in a number of West European countries, and this may be expected to increase. Although there will be significant competition from other sources such as South Africa, American exporters could account for 30 to 50% of the European steam coal import market during the next decade.

In the Pacific Rim, U.S. firms could obtain a lesser share of the steam coal trade, as countries such as Japan, Korea, and Taiwan increase their energy security by diversifying supply sources. The recent surge in demand for steam coal has included the cement industry in Japan, where higher sulfur coal can be effectively utilized. Although Australia and South Africa are geographically better positioned, it is expected that the United States will increase exports to this region.

U.S. COAL EXPORTERS

While the outlook for increased world coal use is potentially bright, I would like to caution that hypothetical projections do not guarantee that U.S. coal exporters will in fact obtain a substantial portion of the future market. Resolving the existing port capacity constraints, better utilization of our inland waterways system, and upgrading and building new rail transportation links, particularly in the West, will help put the United States in a more competitive position.

Poland, South Africa, Canada, and Australia are the main countries in a position to compete effectively for the world coal market. All have some elements of commercial advantage, as well as special circumstances which could affect their ability to compete on commercial terms.

Unlike the United States, which in 1980 exported about 11% of its total coal production, Australia exports more than half of its current output. Australian coal has traditionally dominated the Pacific Rim market, and has also made important recent inroads in the European market. However, recent labor difficulties, mounting bunker fuel costs in colliers, and distance may diminish the competitiveness of Australian coal in Europe. Overall, however, Australia has the potential to be a formidable competitor in international steam coal trade, particularly in the Pacific.

Poland enjoys a significant transportation advantage for the European market. However, Poland's recent internal problems have reduced exports significantly, and could have an impact on future trade.

South Africa represents a significant element of competition for U.S. firms exporting to Europe and the Pacific Rim, given that country's large resource base and competitive prices.

With significant reserves of coal, Canada is also a potential strong competitor in international steam coal trade. Because the majority of these reserves are located in the western provinces, the most obvious markets for Canada's steam coal are the Pacific Rim countries, which are expected to account for a significant share of Canada's steam coal exports in the 1990s. While Europe is also a potential customer, high transportation costs from Canada's west coast could limit the European trade.

Over the longer-run, we could also expect some additional competition from Colombia, the Soviet Union, and China, if these countries develop their resource potential.

ACTION TO FACILITATE COAL EXPORT TRADE

Our domestic coal industry will have the responsibility for increasing U.S. steam coal exports. The government's role is to contribute by maintaining a stable international framework under which commercial business transactions can develop smoothly. The U.S. government has acted to establish and strengthen that framework by strongly supporting the concept that increased steam coal use and trade are necessary to meet future world energy demand, in both the short and long term. We have supported this position with our trading partners both in the International Energy Agency (IEA) and in the course of the annual Economic Summits.

In 1979, the IEA agreement on principles for action on coal was developed to stimulate coal production, use, and trade. In May 1979, the United States was instrumental in creating the IEA Coal Industry Advisory Board. The Board is composed of senior-level industrial officials, who advise member governments on steps needed to facilitate steam coal use and trade.

In addition, we have invited many of our allies and friends in Europe and Asia to visit the United States for discussions on expanded coal trade. In the past year, the National Coal Association and the U.S. government have hosted official coal delegations from France, Spain, the Republic of Korea, Italy, the Netherlands, Austria, and Denmark; there have also been numerous other visits from foreign organizations involved in coal trade. These visits have not only provided a useful forum for the exchange of information, but have also resulted in commercial contracts to export U.S. coal. We believe that these statements and actions provide a sufficient and secure basis under which commercial transactions can proceed.

ACTION BY U.S. INDUSTRY AND FOREIGN PURCHASERS

During bilateral and multilateral meetings dealing with coal exports, we have emphasized that the responsibility for increasing coal trade should not rest solely or even principally with the producing countries; rather, this responsibility should in fact be shared equitably with the importing countries. It is in the interest of foreign coal purchasers to sign long-term contracts to ensure the large capital investment required to develop new mines, improve rail and barge transportation, and expand or construct new export facilities. Prospective coal-importing countries could also encourage direct investment in upstream production facilities, and act to remove barriers to increased transportation and use of steam coal within their jurisdictions. We are confident that American business will respond to the challenge as we in government are committed to maintaining a free and stable coal trading environment.

II

The International Demand
for High Sulfur Coal

THE ROLE OF STEAM COAL IN THE REPUBLIC
OF CHINA ON TAIWAN
K. S. CHANG

Energy is the impetus to social and economic development and its adequate supply is also essential to national security. The Republic of China (Taiwan) has been one of the fastest growing nations from an economic standpoint. Its economy continued to improve steadily in 1980 despite the influences of adverse international economic developments and oil prices. The country's economic growth rate in 1980 reached 6.66%. The gross national product amounted to 40.26 billion U.S. dollars, resulting in an increase of 24.47% over the previous year. The national income this year was 37.07 billion, which was an increase of 24.4%. The country's per capita income reached $2,101, a 22.03% increase over 1979.

The two-way trade for the country reached 39.50 billion in 1980, an increase of 27.9% over the previous year, with a favorable balance of 46 million. The index of industrial production in 1980 reached 8%. This increased production can be traced to the greater emphasis on technology and capital-intensive industries in the Republic of China.

In 1980, the total energy consumption, except 815,000 kl of oil equivalent for exports, was 32 million kl of oil equivalent, resulting in an increase of 6.89% over the previous year. Of the total, 2 million kl of oil equivalent, or 89.37%, were consumed as energy, which was a 6.84% increase over 1979. To classify by forms of energy, coal and its products contributed 7.93%, which was an 11.71% increase over 1979; petroleum products provided 42.7%, which was a 6.57% increase; natural gas gave 3.47%, which was a 1.26% increase; electricity constituted 35.27%, marking an increase of 6.7%.

ENERGY POLICY

In order to guide the management and development of the energy industries to be compatible with the overall economic objectives, our government promulgated early in 1973 a comprehensive energy policy. However, events which have developed since then call for a revision of the policy, with emphasis on accelerated development of indigenous energy, strengthening energy conservation, and further diversification of energy imports. The principal objectives of the energy policy are:

1. to assure adequate and reliable supplies of the various forms of energy needed to sustain economic growth;
2. to enforce energy-saving measures so as to enhance the effects of energy conservation;
3. to strengthen the exploration and exploitation of indigenous energy and to diversify the kinds and sources of energy imports so as to reduce the reliance on petroleum and to alleviate the impact of the possible shortage of oil supply on economic growth;
4. to assure reasonable energy prices and to encourage the reduction of the cost of energy; and,
5. to encourage the use of nonconventional sources of energy such as geothermal, wind, solar, and biomass.

TAIWAN POWER SYSTEM

Existing System

At the end of 1980, the Taiwan Power System had a total installed capacity of 9,056 Mw, of which 15% or 1,386 Mw was hydro, 60% or 5,418 Mw was oil, 980 Mw or 11% was coal and 1,272 Mw or 14% was nuclear. The peak load in 1980 was 6,703 Mw, a 10.4% increase over the previous record. The average load was 4,646 Mw, a 7.4% increase over the previous year. The highest daily average load reached a record of 5,658 Mw.

Production

Net energy production reached 40,813 Gwh during 1980, a 7.7% increase over the previous year. Hydro generation reached 2,905 Gwh, accounting for only 7.1% of the total, and a 36.1% decrease over the previous year. Thermal generation reached 30,096 Gwh, which was 73.7% of total generation, showing a 10.1% increase over the previous year. Nuclear power generation reached 7,812 Gwh, 19.2% of total generation, showing a 30.1% increase over the previous year.

Load Forecast

In conjunction with the nation's economic development, the state- owned Taiwan Power Company (Taipower), having the responsibility of supplying electricity to all customers, has completed the long-term load forecast for the next 15 years. The GNP growth rate as well as the load growth rate from 1980 through 1995 are shown in Table 1.

Expansion of Power System

In order to catch up with the ever-increasing power demand and the nation's economic development, Taipower has pursued a policy of energy supply diversification and cost reduction. Coal and nuclear are the alternative energies to oil for future power generation. According to the latest long-range power development program, the total installed capacity for 1995 will be 33,619 Mw, including 14% or 4,670 Mw of hydro, 11% or 3,755 Mw of oil, 36% or 12,250 Mw of coal, and 39% or 12,944 Mw of nuclear. The system-installed capacity as well as the percentage of various generating capacities from 1981 through 1995 are shown in Table 2, from which it can be seen that emphasis will be put on development of both coal-fired and nuclear plants. Table 3 shows the component variation of energy generation from 1981 through 1995.

REQUIREMENT OF STEAM COAL

Steam Coal Demand for Electricity Generation

The requirement of steam coal for the existing and additional coal-fired units is projected by using a power system operation simulation model with due considerations given to the quality and availability of coal, the characteristics of generating facilities, the optimization of system operation, etc. Provisions for a transportation loss of 4% of total consumption and a stock change for maintaining a five-month inventory are also included in the projection. The specifications for steam coal currently used for electricity generation are shown in Table 4.

Steam Coal Demand for Industries

The increasingly unreliable supply and spiralling prices of imported oil have also encouraged those industries which may use coal as fuel to convert from oil and gas to coal. In 1980, the cement industry, the second largest steam coal consumer (next to electric utilities) in Taiwan, consumed .81 million tons of steam coal which represented 35% of the total fuel requirement for industries, while the other industries, mainly plate glass, ceramics, chemical, coking, etc., consumed 1.54 million tons.

A general survey of steam coal demand for industries was made by means of individual interviews with private industries. The projected future demand of steam coal for industries depends mainly upon their production plans and energy conservation programs.

Table 1
FORECAST LOAD GROWTH (1980-1995)

	1980	1981	1982	1983
Peak load (Mw)	6,703	7,529	8,429	3,388
Growth rate (%)	10.4	12.3	12.0	11.4
Average load (Mw)	4,646	5,251	5,806	6,401
Growth rate (%)	7.4	13.0	10.6	10.2
GNP growth rate (%)	6.7	8.0	8.0	8.0

	1984	1985	1986	1987
Peak load (Mw)	10,417	11,470	12,606	13,807
Growth rate (%)	11.0	10.1	9.9	9.5
Average load (Mw)	7,047	7,716	8,450	9,229
Growth rate (%)	10.1	9.5	9.6	9.2
GNP growth rate (%)	8.0	7.8	7.8	7.8

(Table 1 continues on next page.)

Table 1 (Cont.)
FORECAST LOAD GROWTH (1980-1995)

	1988	1989	1990	1991	1992
Peak load (Mw)	15,105	16,502	18,012	19,656	21,407
Growth rate (%)	9.4	9.2	9.2	9.1	8.9
Average load (Mw)	10,083	11,012	12,045	13,130	14,276
Growth rate (%)	9.3	9.3	9.8	9.4	8.7
GNP growth rate (%)	7.8	7.8	7.5	7.5	7.5

	1993	1994	1995	1980-1995[a]
Peak load (Mw)	23,309	25,376	27,476	
Growth rate (%)	8.9	8.9	8.3	9.86
Average load (Mw)	15,524	16,893	18,246	
Growth rate (%)	8.7	8.8	8.0	9.55
GNP growth rate (%)	7.5	7.5	7.0	7.70

[a] Average Annual Growth Rate

Table 2
SYSTEM INSTALLED CAPACITY

Year	Hydro Mw	Hydro %	Coal Mw	Coal %	Oil Mw	Oil %	Nuclear Mw	Nuclear %	Total Mw
1981	1,426	14	945	9	5,419	54	2,257	23	10,047
1982	1,426	12	1,405	12	5,725	49	3,242	27	11,798
1983	1,493	12	1,905	16	5,725	46	3,242	26	12,365
1984	1,493	11	2,855	21	4,775	36	4,193	32	13,316
1985	2,243	14	4,115	26	4,307	27	5,144	33	15,809
1986	2,688	16	4,665	28	4,247	25	5,144	31	16,744
1987	2,688	15	5,765	32	4,247	24	5,144	29	17,844
1988	2,828	14	7,375	38	4,267	22	5,144	26	19,614
1989	3,476	16	7,925	36	4,267	20	6,044	28	21,712
1990	4,366	19	7,850	33	4,267	18	6,944	30	23,427
1991	4,670	18	8,275	33	4,267	17	8,144	32	25,356
1992	4,670	17	8,825	33	4,267	16	9,344	34	27,106
1993	4,670	16	9,250	32	4,267	15	10,544	37	28,731
1994	4,670	15	10,750	34	4,287	14	11,744	37	31,451
1995	4,670	14	12,250	36	3,755	11	12,944	39	33,619

Table 3
COMPONENT VARIATION OF ENERGY GENERATION

Year	Hydro Gwh	%	Coal Gwh	%	Oil Gwh	%	Nuclear Gwh	%	Total Gwh
1981	4,519	9.8	6,447	14.0	27,233	59.2	7,770	16.9	45,969
1982	4,717	9.3	7,867	15.5	25,177	49.5	13,098	25.8	50,859
1983	4,715	8.4	11,255	20.1	22,569	40.2	17,536	31.3	56,075
1984	5,138	8.3	15,400	24.9	20,656	33.5	20,552	33.3	61,746
1985	5,137	7.6	22,069	32.5	14,978	22.1	25,695	37.9	67,879
1986	5,229	7.0	26,669	35.6	15,194	20.3	27,823	37.1	74,915
1987	6,072	7.4	31,799	38.9	16,065	19.6	27,823	34.0	81,759
1988	6,053	6.8	39,780	44.4	15,084	17.7	27,823	31.1	89,496
1989	6,486	6.6	48,124	49.3	15,172	15.5	27,823	28.5	97,605
1990	6,549	6.1	50,482	47.0	17,640	16.4	32,692	30.5	107,263
1991	7,019	6.0	52,181	44.5	15,727	13.4	42,452	36.2	117,379
1992	7,094	5.6	55,503	43.6	15,870	12.5	48,943	38.4	127,410
1993	7,094	5.1	59,436	43.0	16,385	11.8	55,434	40.1	138,349
1994	7,094	4.7	66,049	43.9	15,271	10.2	61,925	41.2	150,339
1995	7,094	4.4	72,486	44.7	14,193	8.8	68,416	42.2	162,189

Table 4
SPECIFICATIONS FOR STEAM COAL

Item	Spec. Value
Heating value (Kcal/Kg)	6200 Min.
Volatile matter (%)	24 Min.
Ash content (%)	19 Max.
Total sulfur (%)	1.5 Max.
Grindability (H.G.I.)	50 Min.
Ash-softening temperature (°C) (under reducting condition)	1200 Min.
Size (m/m)	50 Max. 2 (30% Max.)
Na_2O in ash (%)	2 Max.

Total Demand for Steam Coal

The anticipated total steam coal demand for both electricity utilities and industry for the next 15 years is tabulated annually (see Table 5). The total demand, subtracting the anticipated supply of domestic coal, is the coal quantity to be imported from worldwide coal sources.

STRATEGY FOR IMPORTATION OF COAL

Diversification of Coal Resources

The remaining coal reserves in Taiwan are estimated to be 200 million tons; the annual production in 1980 was 2.5 million tons, of which only .6 million tons were made available for power generation. To make up the deficiency of coal supply, Taipower imported 3 million tons of coal in 1980 for power generation. Accordingly, steam coal will depend largely on importation. The potential sources of coal for imports include Australia, South Africa, the United States, Canada, Indonesia, India, and European countries. However, the former four countries are considered as the primary sources.

In view of potential uncertainties such as strikes in the mines and loading ports, insufficient berths and loading capacity, lead time needed to develop mines, insufficient inland and ocean transportation, and national protectionism of natural resources, etc., diversification of coal supply from several countries is significantly necessary where a large coal importation program is involved. Therefore, spreading of purchases to several countries provides a measure of protection in case of disruption of deliveries from any one country. Aside from diversification of coal sources, Taipower intends to purchase coal either on the basis of long-term contract or in the form of joint venture on development of coal mines so as to ensure coal quality and supply.

Ocean Transportation

In accordance with a national policy of Chinese ships for Chinese cargo, Taipower plans to build a major portion of the coal fleet required for ocean transportation. Three 125,000 DWT and six 63,000 DWT coal carriers have been contracted to the China Ship-building Company and will be made available to Taipower by 1985. More coal carriers will be built by stages in accordance with future coal requirements.

Domestic Transportation

Cost for domestic transportation is one of the major factors affecting the fuel cost. According to local conditions and economic

Table 5
PROJECTIONS OF STEAM COAL DEMAND IN TAIWAN

Year	Electric Utility	Industry	Total	Domestic	Imports
1981	2.9	3.0	5.9	2.6	3.3
1982	3.7	3.8	7.5	2.6	4.9
1983	5.2	4.6	9.8	2.6	7.2
1984	8.0	5.5	13.5	2.6	10.9
1985	10.7	6.3	16.6	2.6	14.0
1986	12.0	6.3	18.3	2.6	16.7
1987	14.3	6.8	21.1	2.6	18.5
1988	17.8	7.3	25.1	2.6	22.5
1989	21.1	7.8	28.9	2.6	26.3
1990	21.2	8.3	29.5	2.6	26.9
1991	21.7	8.8	30.5	2.6	27.9
1992	23.2	9.5	32.7	2.6	30.1
1993	24.8	10.1	34.9	2.6	32.3
1994	28.1	10.7	38.8	2.6	36.2
1995	30.9	11.4	42.3	2.6	39.7

considerations, the domestic transportation system can be classified into three types according to the distance between the unloading terminal and the power plant, as given in Table 6.

Table 6
DOMESTIC TRANSPORTATION SYSTEMS

Distance Between Unloading Terminal and Power Plant	Adopted Facility
Nearby (10 km)	Conveyor belt system
Short distance (10 km and 35 km)	Truck/barge system
Long distance (35 km)	Railway

Blending of Coals

In view of the limited availability of low sulfur coals and the successive addition of coal-fired units to the power system, blended coal would be considered for power generation. Coals can be well blended and successfully fired in any new unit that has been designed to fire such fuel. However, the complete knowledge of the quality of coal that is to be blended, the method and extent of blending, and the expected range of properties of the blended coal all have to be investigated before the design can be completed. Since the performance of blended coals cannot be entirely predicted by simple average of the properties or constituents of the makeup coal, a laboratory test is required in order to obtain the properties of the resultant blended coal.

According to the preliminary results of laboratory tests for blending low sulfur coal from western sources with high sulfur coal from eastern sources at an 80:20 weight basis, the quality of such blended coal could meet the current specifications for imported steam coal. However, as far as sulfur content is concerned, it should not exceed 3%.

Contracted and Negotiated Coals

As described previously, Taipower, being the largest user of steam coal, has to import annually a large quantity of steam coal for electricity generation. In order to secure sufficient coal from abroad, Taipower has been actively seeking coal sources. To date, a total of 92 million metric tons for the 1981-1995 period has been contracted; negotiating coal through coal sources is summarized in Table 7.

Table 7

CONTRACTED OR NEGOTIATING COAL

Year	Coal Amount (1,000 tons) A[1]	Coal Amount (1,000 tons) B[2]	U.S.A. A	U.S.A. B	Australia A	Australia B	South Africa A	South Africa B	Canada A	Canada B
1981	3,334	150	65.5	-	18.0	-	16.5	100.0	-	-
1982	3,345	600	65.6	-	19.4	50.0	15.0	50.0	-	-
1983	4,510	900	61.2	-	18.9	33.3	15.5	66.7	4.1	-
1984	5,470	1,400	58.0	-	16.5	57.1	16.4	42.9	9.1	-
1985	6,970	1,350	48.4	-	17.2	77.8	12.9	22.2	21.5	-
1986	8,800	1,350	44.3	-	15.9	77.8	5.7	22.2	34.1	-
1987	9,300	1,400	47.3	-	15.1	78.6	5.4	21.4	32.2	-
1988	9,400	1,400	46.8	-	16.0	78.6	5.3	21.4	31.9	-
1989	9,900	1,400	49.5	-	15.1	78.6	5.1	21.4	30.3	-
1990	7,300	1,400	31.5	-	13.7	78.6	6.8	21.4	48.0	-
1991	5,000	1,400	20.0	-	10.0	78.6	-	21.4	70.0	-
1992	5,000	1,400	20.0	-	10.0	78.6	-	21.4	70.0	-
1993	5,000	1,100	20.0	-	10.0	100.0	-	-	70.0	-
1994	5,000	1,100	20.0	-	10.0	100.0	-	-	70.0	-
1995	4,000	1,100	-	-	12.5	100.0	-	-	87.5	-
Total	92,329	17,450	40.7	-	14.6	76.8	6.6	23.2	38.1	-

[1] Contracted coal
[2] Negotiating coal

Ambient Air Quality Standards

The Government of the Republic of China promulgated the Air Pollution Control Act early in May, 1975. This Act is enacted in order to prevent air pollution and to protect public health. The National Health Administration is charged with the responsibility for executing this Act. The Ambient Air Quality standards of this Act are presented in Table 8.

Table 8
AMBIENT AIR QUALITY STANDARDS

Item	Interval	Ordinary Area	Industrial Area
1. Particulates			
A. Suspended particulates (less than 10 um particle diameter)	Monthly average	210 ug/Nm^3	240 ug/Nm^3
	Annual average	140 ug/Nm^3	160 ug/Nm^3
B. Dust-fall (greater than 10 ug particle diameter)	Monthly average	260 ug/Nm^3	290 ug/Nm^3
	Annual average	170 ug/Nm^3	190 ug/Nm^3
2. SO_x			
A. Concentration value	1 hour average	0.3 ppm	0.5 ppm
	24 hour average	0.1 ppm	0.15 ppm
	Annual arithmetic mean	0.05 ppm	0.075 ppm
B. Accumulative value	$100cm^2$/30 days	50 mg (SO_3)	90 mg (SO_3)
3. NO_x			
A. Concentration value	1 hour average	0.05 ppm	0.1 ppm

B. Limitation: allowed 36 days exceeding the limited value

4. Stack emission rate

A. SO_x: 2000 ppm

B. Particulates: 700 ug/Nm^3

C. Smoke concentration:
 (1) Ordinary condition limited to Ringelman No. 2
 (2) Initial burning stage limited to Ringelman No. 3
    but does not exceed an accumulative time of
    3 minutes within 1-hour duration

THE ROLE OF IMPORTED STEAM COAL IN SPAIN
JOSE SIERRA-LOPEZ

THE SPANISH ENERGY PROBLEM

Spain is now in the process of building up a parliamentary and democratic political system, a task that would have been much easier within a framework of economic prosperity, but the effort of stabilizing our democratic system comes along with the need to readjust our economy in the midst of an historical economic crisis that affects all the nations of the world. One of the most serious challenges we have to meet is how to decrease our dependence on imported oil without seriously impairing the wealth and prosperity of our country. Some figures will be helpful to understand the present situation of Spain from the viewpoint of energy supply and demand.

Spain, with a population of 37 million inhabitants and an energy consumption of 78 million Btu per person per year, is very far away from self-sufficiency because of the very important role played by oil, as the data in Table 1 illustrate; domestic wells supply only 3.5% of oil demand and, although Spain exports some refined products, our oil trade balance showed a deficit of 11 billion dollars in 1980, about 6% of our gross industrial product.

Table 1
PRIMARY ENERGY CONSUMPTION IN SPAIN IN 1980

	Quads	M. TOE	Percentage
Hydropower	.27	6,804	10.4
Nuclear	.04	1,079	1.6
Natural gas	.07	1,758	2.7
Coal	.54	13,534	20.6
Oil	1.68	42,376	64.7
	2.60	65,551	100.0

In relation to the world energy resources, Spain can consider itself well endowed with uranium but not with fossil fuels; its fossil fuel resources represent about 3% of the world's total, while its gross domestic consumption represents about 1%. In Spain, as in most countries, coal was the major industrial development component, especially around the 1940s, but the increase in coal use ended during the 1950s. The substitution process was more drastic in Spain than in other European nations, and therefore, there is no other country in Europe--except Italy--as dependent on oil as Spain is.

In 1950, solid fuels were the most important energy source, accounting for almost three-fourths of the total consumption, followed by hydro. Oil did not even account for 10% of that total. Coal's participation came down to 16% in 1978 from 74% in 1950, and although energy demand more than doubled, coal use declined sharply during this period. However, the adjustment to the impact of the energy crises of 1973 and 1979 gave way to the recovery of the coal sector, and coal production increased from 14.2 metric tons in 1975 to 27 metric tons in 1980. Coal's share of the total primary energy consumption in 1980 was 20.7%, but oil was 64.6%, and hydro was only 10.4%, although under medium conditions hydroelectric power could have taken three more points away from oil.

The Spanish energy problem is, as in most Western nations, how to shift from oil to alternative energy sources. The National Energy Plan for the 1978-1987 period foresees that oil will not even account for 50% of the total primary energy required during 1987. Toward this goal, the plan is to develop a broad nuclear power plant construction program, utilize the existing hydro resources to their maximum, increase national energy production and coal and gas consumption, and, in addition to all feasible energy conservation measures, develop the use of new energy sources, especially solar. This way, it is expected that of the total 3.7 quads to be required in 1985, coal will supply 23%, oil 49%, nuclear 11%, hydro 10%, and gas 7%.

STEAM COAL NEEDS IN SPAIN

A structure of primary energy sources such as the one indicated above calls for extensive use of steam coal, which will be influenced by the following factors: scarce oil supplies and their price, changes in the time required to put nuclear power plants in service, annual hydroelectric energy production, the country's energy needs, and the share of the market that coal can take in accordance with its price as compared with the price of oil.

Spain's future steam coal demand will basically come from three sectors: electricity production, industry needs, and residential/commercial users.

Electricity Production

The current thermal power plant construction program is aimed at the addition of 6,862 Mw during the 1980-1990 period, as well as the conversion from oil to coal of 2,141 Mw currently installed. Taking into account the installed capacity at the end of 1980 (about 6,500 Mw) and the new plants to be built, coal would be used to generate 7,300 Mw in 1983, 11,800 Mw in 1985, and 12,300 Mw in 1987.

Industry Needs

This includes an ongoing program for reconversion of the cement industry and will use about 2.7 million metric tons of coal by 1982. Other studies are underway regarding the reconversion of the sugar and brick industries, which would use about 1 million metric tons per year.

Residential/Commercial Users

Homes and non-residential, non-industrial users consumed 1.3 million metric tons of coal in 1980, and they are expected to use increasing amounts in the future in accordance with coal competitiveness and the impact of environmental protection regulations.

SPANISH DEMAND FOR IMPORTED STEAM COAL

To meet the short- and medium-term needs of steam and metallurgical coal, the Spanish Coal Production Plan seeks to increase coal production from the 12 million tce, and 15.4 million tce in 1977 and 1980, respectively, to 18 million in 1985 and 20.6 million in 1987, but most of the additional production will be made of lignite. These increases, however, are not enough to supply the expected demand growth; therefore, imports of steam coal will increase substantially, although the exact amounts to be imported yearly will depend on the evolution of domestic demand and production. So far, only the electricity producing and cement manufacturing sectors have been considered in the assessment of imported coal needs. It is expected that domestic production will be sufficient to meet the demands from other sectors.

In the electricity generation sector, imported coal will either supply power plants that are being built for burning it, or will compensate for deficiencies in quality or quantity in plants burning domestic coal. On the other hand, about 70% of existing cement manufacturing plants will shift from oil to imported coal because of their location; other plants will burn coal from mines in their area.

A total of about 10.6 million tce of steam coal will be imported in 1985 and distributed in the following way among the main consuming geographical regions:

1. Andalusia, 42%;
2. Catalonia and Aragon, 20%;
3. NW (Galicia, Asturias, Leon, and Palentia), 32%.

The proportion will change slightly in 1987: 40, 25 and 25, with a total consumption of about 13-14 million tce.

COAL IMPORTS POLICY IN SPAIN

Objectives and Directives

Within the framework of its energy policy, Spain intends to reduce and diversify its dependence on oil, substituting part of it for imported coal. The strategic character and the impact of this shift on the national economy requires completing it under the best security and overall price conditions. Due to the fact that sea freight affects the final cost in great proportion, it is necessary to design consistent marine transportation and port construction policies, along with a long-term policy of investments in coal production abroad.

The attainment of the above mentioned objectives will evidently be influenced by the existence of supply management agencies with size and capabilities big enough to provide the required amounts of coal within a free-market framework. Therefore, it is estimated that the minimum size needed to conciliate the diversification of sources and capital investment criteria must be 5 million metric tons in 1985.

Imports

If the criteria for diversified supply sources and contract modalities are considered, a reasonable objective for 1990 could be the following:

1. approximately 40% of imported coal should come from mines owned by corporations in partnership with Spanish companies linked to long-term supply contracts for coal amounts exceeding current Spanish participation;
2. about 40% more should come through long-term contracts, some of which could be linked to the above-mentioned participants; and
3. the remaining 20% would be available for short-term and spot purchases, or for financial-industrial, long-term agreements payable with coal supplies.

In general terms, we feel that the long-term success of any agreement must be based on mutual confidence. We recognize that producers need to have a guarantee of the sale of their commodities at prices allowing a reasonable benefit over their costs, but we need long-term guaranteed supply and price stability.

Ocean Transportation

Spanish coal purchases will be made on an FOB basis; exporters will take care of freight management by using Spanish or foreign flagships under free-market conditions. In order to reduce shipping costs to a minimum, 125,000-150,000 DWT vessels are needed, up to a total of 3.5 million DWT in 1987, to handle the 17 million metric tons of coal that will be imported.

Ports and Facilities

To receive the coal to be imported by 1985, new terminals will have to be built in Spain for the following fundamental reasons:
1. current capacity of Spanish ports will not allow the handling of more than 4.5 million metric tons of imported steam coal per year; and
2. the agreements with producing countries under relatively economical conditions imply the use of ships of 125,000-150,000 DWT, and existing Spanish ports are not deep enough.

Terminals will be built at the following locations:
1. Algeciras will receive ships of more than 150,000 DWT; it will handle more than a million metric tons per year and a storage area will also be built;
2. in Gijon, the town council is financing the construction of a new terminal with dredging for ships of 150,000 DWT; unloading equipment will be installed to handle more than 12 million metric tons, 5 million of which will be steam coal; and
3. in Carboneras a port to receive ships of 60,000 DWT will be built, with an unloading capacity of almost 3 million metric tons per year.

The investment required for the construction of these installations is estimated at 21,000 million pesetas (1980 value). The installations are expected to start operations in 1984.

SOURCES OF SPANISH COAL IMPORTS

Decisions on the origin of imported coal will depend on the following factors: distance to coal producing countries, their own export capabilities, the availability of new ports both in Spain and abroad, and a stepwise increase of Spanish demand, as well as supply security.

All things considered, the United States and South Africa will be the main suppliers in the next few years, while countries such as Australia, Canada, Poland, Colombia, and China will become more and

more important in the following years. Up to 1984, we will depend
on the United States for 60 to 80% of our coal supplies. Then, the
percentage will be reduced to 50 to 40% in favor of other countries
such as Australia (18-19%), Canada (5-13%), South Africa (14-12%),
Poland, the Soviet Union, China, and Colombia.

SPANISH IMPORTING AGENTS: INI AND CARBOEX

With the purpose of becoming an efficient agent capable of providing 40 to 50% of the total Spanish steam coal needs in a
free-market environment, the National Institute of Industry (INI)
created CARBOEX (Sociedad Espanola de Carbon Exterior), which was
incorporated on March 20, 1980. In accordance with the above,
CARBOEX expects to obtain a market for 6.5 million metric tons in
1985, and 11.1 million metric tons in 1990. Its main objectives are
as follows:
1.   to link 40% of the imports to long-term contracts;
2.   to gain minority participation in joint ventures overseas,
     securing an additional 40%; and,
3.   to insure competitiveness of CARBOEX supplies through capital
     investments in the ports of Algeciras, Gijon and Carboneras,
     and in other smaller terminals.

COAL QUALITY REQUIREMENTS IN SPAIN

Table 2 lists the specifications of the coal needed by the
cement manufacturing and electricity sectors:

Table 2
COAL SPECIFICATIONS

Characteristics	Cement	Utilities
Moisture	8% max.	8% max.
Ash	15% max.	13% max.
Volatile	18-30%	24% min.
Caloric value	6,200 Kcal/Kgr. min.	6,500 Kcal/Kgr. min.
Sulfur	2.5% max.	1.5% max.
Hardgrove	55 min.	55 min.
Ash fusion	1,350 Gr. C. max.	1,350 Gr. C. min.
Size	0-55 mm.	0-55 mm.

Current Spanish regulations (Decree 833/1975 of February 6, Annex 4, Section 1.1) establish the following emission standards applicable to thermal power plants burning coal with a maximum of 1.5% sulfur and 20% ash:

1. Solid particulate   150 mg/Nm^3
2. SO_2 2,400 mg/Nm^3

HIGH SULFUR STEAM COAL IN FRANCE
WILLEM G. M. DANIELS
JEAN FAUCOUNAU

What is high sulfur steam coal for France? The sulfur content of the French domestic coals, being a maximum 1.8%, it is not surprising to note that, for the French end-users, high sulfur coal is any coal of 11,000 Btu exceeding 1.8% sulfur which means 3.3 lbs. of SO_2 per one million Btu. This is more or less the present yardstick. However, in a number of limited cases, this ceiling is being relaxed, but only on the basis of a Btu-related content and in no case above the 2% level for 12,000 Btu, i.e., again the 3.3 lbs. of sulfur per million Btu.

FRENCH UTILITIES AND BUSINESSES

Since no scrubbers are installed (but only ash precipitators), the utilities adhere very strictly to the 1.8% ceiling although stringent regulations, strangely enough, do not exist. The utilities are, therefore, buying under a discipline of self-constraint. For 1981, we today have purchased about 7 million tons.

As for cement mills (general industry and urban heating), the bracket these consumers are looking for is 1.5 to 1.8%, but some of them are asking for a sulfur content as low as 1% for technical reasons. For 1981 we have about 1 million tons of coal under contract for this category of clients.

FUTURE COAL CONSUMPTION IN FRANCE

For utilities, no change is expected in the basic requirements; however, we are currently looking into the possibility of buying mixes (we would not really call them blends) of high and low sulfur coals, provided the volatile content of each component does not differ by more than 2 to 3%. This requirement is of great importance since a larger difference of volatiles in a mix which is not strictly homogenous may create serious combustion problems. We are presently studying such possibilities out of New Orleans, but it must be realized that such a mix would certainly not have the homogeneity of an East Coast blend.

For the moment, we can expect a change of attitude to come from the cement works. I think that in September, 1981, we may know more about the possibility of burning coals with 2.3 to 2.5% content. The point is that the cement mills have converted themselves from the humid to the dry process in order to save energy. Under the old, humid process 3 to 3.5% sulfur was not a problem.

FLUIDIZED BED COMBUSTION IN FRANCE

Concerning fluidized bed combustion, so far not much has been done on the subject with today's availability of French and foreign coals with a low sulfur content. However, one maker already markets a high temperature fluidized bed boiler. But in France, as elsewhere in the world, there are no fluidized bed boilers in operation with low temperatures. I am glad to report that a leading boiler constructor has just requested us to ship as soon as possible a 4-ton sample in order to carry out tests under atmospheric pressure.

SYNFUELS IN FRANCE

The top limit of sulfur will be much higher (3 to 4%) when using coal for gasification or methanol purposes. There are a number of projects under consideration. The one which is currently mentioned is the Gaz de France gasification pilot plant which would use 1000 tons of coal per day. If this proves successful, other units might be built and require a total of 3 million tons of coal by 1990. Regarding methanol plants either in France or to be built in the U.S. by French interests, I am unfortunately not in a position to give any information since these projects are still at a very early planning stage.

JAPANESE COAL USE AND THE
INTERNATIONAL COMMUNITY
AKIRO KINOSHITA

Coal could provide the trigger to restore the economic
mechanism and supply stability to the field of energy because it
gives free and competitive options which come from the availability
of so many types of coal and sources of supply of this widely
distributed and abundant resource. These options should be made
viable through vigorous technological innovations at all stages of
the coal chain.

It is a global challenge to break the world's economic
stagnation and associated energy constraints. Coal will bring
various economic benefits, including the improvement of
international balance of payment situations through a coordinated
buildup of world trade, reorganization of large-scale
infrastructures, development of related technologies and the
creation of direct and indirect effective demands through
stimulation of the coal-related industries. A stable and long-term
coal trade can become an effective means to cope with the
international trade imbalance and its instability, which is mainly
attributable to the unstable oil situation.

At the same time, the creation of options for abundant and
stable substitute energy resources based on international
cooperation has a great significance for opportunity costs to
prevent the collapse of the established world economic order and
possible conflicts between industrialized countries in securing
energy resources.

Increased coal use requires comprehensive international
coordination and management for the consistent and effective
operation of an overall chain. Each component of this chain is
linked with the respective regional system, consisting of a series
of facilities such as those for mining, processing, transportation,
shipping, maritime transportation, storage, distribution,
environmental protection, combustion and waste disposal.

POTENTIAL COAL DEMANDS IN JAPAN

The major steam coal use in the future of Japan, as in other
countries, is projected to be in electricity generation, which is

forecast to consume more than 70% of total steam coal by 1990. Estimates of coal requirements in the electric market are strongly influenced by assumptions about the rate of growth of electricity demand and the expansion of nuclear power.

In this projection, electricity is assumed to increase more or less at the rate of economic growth, taking into account progress in energy conservation and the continued transformation of the industrial structure. Electricity/GNP elasticities in the major industrialized countries show 1.5 to 2.0%. These are assumed to shift, in common, to electricity in the energy consumption pattern. A 4.5% annual rate of economic growth is seen between 1980 and 2000.

Japan is one of the most active countries in seeking to realize international agreements and principles for shifting from oil to coal, with special consideration for a balanced approach which includes nuclear power development to overcome its particularly serious energy situation. At present, eight coal-fired power plants, 8,325 Mw, are under construction or preparation. In addition, five plants, 5,000 Mw, are to be approved in the near future as well as the coal conversion of four oil-fired plants, 1,149 Mw. According to the long-term development program of electric utilities, installed capacity of coal-fired plants is to reach 28 GW by 1990.

Nuclear power already plays a substantial role in electricity generation. It supplied 16% of the total electric energy in 1980 and is expected to increase its share up to 31% in 1990, according to the target set by the government. However, there are several predictions as to whether it is possible to install all this nuclear capacity for the future. Exxon forecasts 37 GW, the Institute of Energy Economics 36 GW and the U.S. Interagency Coal Export Task Force 20 GW-27 GW, compared with the government target, 51 GW-53 GW, respectively, for 1990. Also, Exxon forecasts 75 GW, ICE 54 GW-68 GW compared with WOCOL, 90 GW-100 GW, respectively, for 2000.

On the assumption of 36 GW for 1990 and 70 GW for 2000 with respect to nuclear capacity, Japan's needs for steam coal would be arithmetically calculated to be 89 Mtce for 1990 and 192 Mtce for 2000, including requirements for other industries. Japan will then have to import 171 Mtce for 1990 and 276 Mtce for 2000, including metallurgical coal. Feedstock requirements for synthetic fuel are not considered in these figures.

Of course, import-export trade regional patterns will be decided by factors such as economics, security of supplies, the condition of the infrastructure, viability of competitiveness, government intervention and so forth. Roughly one-third of the total import requirements would depend upon the United States. In this sense, it is no doubt very urgent to set up an overall scheme for coal transactions between the two countries.

ENVIRONMENTAL ISSUES IN JAPAN

Japan has been regarded as the most advanced country in developing environmental control. The Electric Power Development Company (EPDC) has been operating limestone type flue gas desulfurization units since 1967 for its coal-fired power plants with an average service factor of almost 100%. The company has also been developing dry-type FGD, a pilot plant which has been in service since 1978 and which is intended for commercialization by 1982.

Most of the environmental risks from coal use are amenable to technological control. On the other hand, each increment to reduction increases the cost. The SO_2 control costs are estimated at \$16 per ton coal for 100 ppm emission and \$18 for 50 ppm emission. In the case where coal with a high sulfur content of more than around 3% is used, the costs jump discretely because dual FGD units must be prepared to meet the existing environmental requirements. In addition, it would not be easy to negotiate with local governments for change to using coal with a higher sulfur because of adverse public acceptance, even though emission limits can be cleared. At present, in almost all cases, in Japan 1% of sulfur content in terms of weight would be a parameter as the upper limit generally adopted, even on the assumption of providing FGD.

Environmental problems are highly regional. The control measures differ by countries because of geographical, climatic, ecological, industrial circumstances, and social perceptions of values. Almost all local governments have independently enacted pollution control ordinances which are much more stringent compared with the national standards. The important thing is not to consider loosening the standards and regulations but to overcome the environmental problems through technological improvement in order to effectively realize the coal objectives based on public acceptance.

ILLINOIS COAL IN JAPAN

Because of the high portion of inland and ocean transportation costs to the total delivered cost for Japan, the coal market is, in general, considered to be geographically bounded. Of course, the size of this geographical market area is determined by the delivered cost of competing coals, which is determined by a combination of production costs including capital charges, royalties and taxes, transportation costs, utilization costs (including pollution control), and the degree of market price competition. In particular, we believe it is important to bring up a variety of viable options and alternatives in order to fulfill the function of market competition.

In the case of coal from the Illinois Basin to Japan, we can point out the advantages of relatively stable, low costs for inland waterway transportation as well as high Btu content against the disadvantages of high ocean transportation costs and high sulfur content in comparison to western coal.

A larger ship will be required to increase the economies of scale in order to overcome the handicap of maritime transportation. These would be required to meet at least 150,000 DWT by 1990. Maritime freight costs may be saved by approximately 40% in the case of 150,000 DWT size compared with that of 60,000 DWT. With respect to maritime transportation costs for Japan, the important viewpoint is to compare alternatives between passage through the Panama Canal and the use of 100,000 DWT from the West to Japan. The cost difference would appear to reach around $17 per ton. In the case of the route for Japan, it would be limited to compensation for the handicap of distance even with ships larger than 150,000 DWT, though there is reason to expect some reduction of the present cost through the Panama Canal by means of the use of large super carriers. Furthermore, if a new Panama Canal, which could accept more than 100,000 DWT, were opened, economic viability of coal to be shipped from the Gulf Coast to Japan would be greatly increased. Also, there could be the application of coal-fired ships on the assumption that the price of bunker oil for the future will increase, though the price of bunker oil is slack at present.

With respect to the handicap of a high sulfur coal with content higher than around 3%, a drastic reduction of production costs through the introduction of innovative production systems must be expected in order to be able to compete with western coal with low sulfur content based on large scale surface mine systems.

It is expected that there is sufficient possibility for innovation in coal chain technologies involving wide aspects in operation management since, historically, priority has never been given to coal development, and also the industry itself has been in decline in contrast to the flourishing oil industry. I am convinced that a comprehensive, innovative approach to the overall Illinois coal chain is an essential key to establishing an effective and efficient option in the Pacific coal market.

In addition, more attention should be paid to the characteristics of this supply region in terms of stability of transportation systems based on an independent and competitive system which may contribute to an emergency. And also, relatively competitive supply systems along the Mississippi, which can involve various sources in this region as well as competitiveness of waterway transport enterprises, would be noted.

ACTIONS TO INCREASE COAL EXPORTS

Finally, I would like to point out several measures related to viability of coal to be shipped from the Gulf Coast; for this, we need to do the following:

1. clarify the overall Mississippi coal chain system, including the subsystem from mine to waterway, opening for international buyers markets which will in the future involve large, international regional coal centers;

2. extend public support for the Mississippi coal chain in order to stabilize overall transport charges, taking into account repercussive, external economic effects induced by coal development in this region;

3. encourage technological innovation in the waterway transportation system linked with bulky, ocean coal transportation, including a continuous rapid transshipment system;

4. encourage innovation in the coal mines, including modernization and integration of small mines through financing, taxation, a plant lease system, and so forth;

5. avoid the increase of imposed taxes and public charges to intensify competitiveness and government interventions and inefficient procedures on international trade and investments for coal development;

6. organize a coal information system related to reserves, quality, production, and the transportation systems in this region in order to introduce the effective supplier into the free and competitive market;

7. study the feasibility of establishing a large-scale coal stockpile on the lower Mississippi on an international basis, taking into account the regional characteristics in terms of stable operations and physical endowments; and

8. promote international cooperation for examining the feasibility and framework related to the a Panama Canal project in consideration of broader global effects.

COAL USE IN KOREA
BONG SUH LEE

In 1980, the total consumption of energy in Korea amounted to 38 million tons. Petroleum accounted for 62%, and anthracite, 26%. Firewood, nuclear, hydro, liquified petroleum, gas, and bituminous coal made up the balance. By 1986, energy consumption is expected to increase to 59 million T.O.E. With this increase, the configuration of energy supply sources will change significantly. Petroleum dependency is expected to decline markedly from the current 62% to 48%, due to rising substitution of alternative sources, such as coal, gas, and nuclear energy.

KOREAN COAL PROJECTIONS

Coal is expected to increase to 43 million metric tons in 1986 from the 1981 level of consumption over 30 million tons. Of the 30 million tons of coal consumed at present, 22 million tons are anthracite, but only 9% of this requirement is met by domestic production. The demand for anthracite for the next 5 years is expected to increase, only at a lower rate. More than two-thirds of the 13 million ton increase in domestic coal in the next five years is attributable to a greater use of bituminous coal, particularly in the field of power generation and cement manufacturing. Furthermore, it is forecast that the steam coal requirement alone in the year 1991 will be around 12 million tons.

COAL IMPORTS

At the present time, we have two coal-fired plant projects under construction. Each project has two 500 Mw plants. Both projects are to be operational by the middle of the 1980s. We have already signed contracts with the companies from Australia and Canada, to supply 4 million tons of coal per year. We are now negotiating for an additional 1 million tons with several firms from Australia, Canada, and the United States. The future outlook for the demand for steam coal is a direct problem for the construction of new coal plants and conversion of the existing plants to coal from oil. However, the non-availability of bituminous coal in Korea requires the import of all new amounts of steam coal, and has required emphasis to be placed on the importing of steam coal from various countries in Europe, and, in the very near future, the United States.

POSSIBLE BARRIERS TO IMPORTS OF U.S. COAL

I would, however, like to express some of the probable barriers we may encounter by importing U.S. coal, particularly high sulfur steam coal. First, the Environmental Protection Regulations in Korea presently impose a regulation that require the emission level of sulfur oxide to be below 1800 ppm. Second, the price of U.S. coal is currently not competitive in the Korean market. In 1981, for instance, coal from Australia and Canada was being negotiated at about $50 per ton FOB basis, but the price of U.S. coal was about $7.00 higher per ton. In addition to this higher cost, layover shipment of this would raise transportation costs by about $3.00 per ton. Third, we are considering, at this time, the implications of entering into any long-term contracts involving the importing of U.S. coal. This issue is mainly due to the possibility that U.S. coal companies could not adequately facilitate the increased coal movement from the U.S. to Korea.

FUTURE PROSPECTS FOR U.S. COAL IN KOREA

Even after considering these possible problems with the importing of U.S. coal, I still maintain optimistic views concerning the utilization of U.S. coal in Korea in the near future. First, I believe that there is ample possibility of utilizing high sulfur coal in Korea. This can be accomplished either by learning the cause of the difference of contents in U.S. coal, or by relaxing the sulfur oxide emission regulation. In the near future, we will study the possibility of relaxing this sort of regulation, particularly on power generation plants located in the eastern zones. Second, although the price of U.S. coal is still very high in the Korean market, as I have mentioned, the gap between U.S. coal prices and other competing coal prices is narrowing. This is due to the rapid increase in prices for both Australian and Canadian coal. More specifically, the U.S. coal prices were approximately 30% higher than Australian and Canadian coal in 1979. Currently, the price differential has been reduced to less than 20%. Third, Korea has been relying mainly on Australia and Canada for importing coal. Korea is now actively searching for U.S. coal for diversifying the source of coal supply for a variety of reasons. This includes uncertainty of availability of coals because of the level of strikes and conditions in both Australia and Canada. If the U.S. can become more aggressive in expanding markets and in maintaining negotiations for importation, I firmly believe that there will be ample possibilities of opening a new coal market in Korea. Today's conference certainly is material proof of such an effort by the United States.

JOINT STUDY BY WESTERN STATES AND KOREA

I may also note a similar effort exerted by the governors of the western states in proposing a joint study in order to explore the possibility of promoting the use of United States coal in the Far East. The Korean government is in full support of this study and has decided in principle to contribute our share to the research fund. We are equally interested in the future possibility of importing U.S. high sulfur coal and will be an active participant to any such agreement that will ensure a stable supply of coal at a reasonable price in our part of the world.

III

The Future of High Sulfur Coal
in the World Energy Market

BACKGROUND AND SUMMARY
CARL BAGGE

The year 1980 set a record for U.S. coal exports as 90 million
tons of steam and metallurgical coal were shipped to Canada and
destinations overseas. The experience of 1980 was a turning point
in the history of U.S. exports. In the past, 8-10% of the coal
which was exported was primarily metallurgical grade coal for use in
the coke ovens of Japan and Europe. Little steam coal was exported
to destinations other than Canada. In 1979, however, an increase in
demand for steam coal for use under utility boilers and for
industrial uses was experienced as other nations began a rapid
substitution of coal for expensive and politically unreliable oil.

In 1978, no steam coal was exported. In 1979, 2.5 million tons
of steam coal were exported, in 1980 16 million tons, and in 1981
approximately 25 million tons of steam coal will be shipped abroad.
Over the next decade United States coals can play an increasingly
important role in meeting the steam coal needs in Europe, the
Pacific Rim nations and in other countries. The Interagency Coal
Export Task Force report forecasts a potential world steam coal
import demand which could be as high as 280 million tons in 1990 and
565 million tons by 2000. This report forecasts that the United
States' most probable share of this market will be 64 million tons
in 1990 and 197 million tons in 2000.

National Coal Association forecasts are slightly higher,
projecting U.S. shipments of up to 79 million tons of steam coal to
overseas destinations in 1990. Couple this with an estimated export
of 55 million tons of metallurgical coal, and 134 million tons of
U.S. coal could be shipped overseas in 1990.

Forecasts are, however, not yet reality, and there are a number
of constraints to be addressed before the United States has the
capability of exporting the quantities which will be needed by
customers overseas in future years.

I mentioned that the 90 millions tons shipped for export in
1980 set a record. However, an additional 10 million tons were not
sold to customers overseas due to constraints imposed by the
capacity of seaport and coastal harbors, problems which existed
because demand for United States coals was so much greater than the
capability of transportation and port system.

Despite extraordinary effort put forth by coal shippers, inland coal carriers, and seaport operators in seeking maximum utilization of coal ports, seaport and harbor congestion, which was evidenced by as many as 150 colliers at anchor in harbors and outer harbor areas, resulted in not meeting 10 to 15 million tons of overseas demand for U.S. coal in 1980. This is an intolerable situation when coal producers can now supply more than 100 million tons above existing production levels using in-place mining capacity and when 20,000 coal miners are waiting for work. The coal that was exported last year created employment for some 32,700 coal miners and an additional 30,000 jobs in support industries as well as adding over 4.5 billion dollars to the balance of payments, but the lost tonnage cost the U.S. at least another half-billion dollars.

HARBOR CAPACITIES & THE OVERSEAS MARKET

To assure that the U.S. position in the world coal marketplace becomes fully established and well secured in the next few years and over the last 2 decades of this century, those involved in the U.S. coal export delivery system must act quickly to upgrade and expand coal ports by the use of ground storage coupled with modern bulk commodity coal-handling equipment and adequate coal pier space for ocean colliers at existing and new coal ports on the East, Gulf, and West Coasts. Already there are several projects underway to provide such improvements at seaports in Pennsylvania, Maryland, Virginia, North Carolina, South Carolina, Georgia, and Alabama. Coal producers in the Midwest will benefit from the improvements now underway in the New Orleans area.

Private enterprise has demonstrated that it can, and will, furnish landside improvements needed to provide additional, modern coal port capacities for U.S. coal exports. The benefits of such initiatives are apparent, and their impact in terms of reduced congestion could begin to be seen in the near future. This can be expected to allow U.S. coal exporters to recapture opportunities lost due to port congestion and inability to load waiting colliers in 1980. Such circumstances led a number of potential customers away from U.S. coal last year. They were discouraged by the long lines of waiting vessels and huge demurrage costs, which raised the price of U.S. coal delivered to Europe by more than $12 a ton--some 15 to 17% of the delivered price.

Although a great deal of satisfaction can be gained from many ongoing and imminent accomplishments, the challenges ahead require even greater effort and active government participation. These are:
1.   dredging seaport channels and harbors to depths necessary to accommodate colliers that are 100 thousand to 150 thousand or

more DWT in size in addition to the 60,000 DWT vessels now commonly used, and

2. adopting a fast-track permitting process for groundside, harbor, and channel improvements at the ports along with an expedited environmental review procedure in order to achieve modern coal port operations on the East, Gulf, and West Coasts and on the Great Lakes.

The case for dredging was made very well by the Interagency Coal Export Task Force Report. It made the point that the U.S. is in a position to capture a major share of the growing world coal market. But cost is a key factor. The delivered price of U.S. coal in Europe can be reduced by $6 a ton--105% of the current price--by capturing economies of scale in transportation. Dredging seaport and coastal harbor channels to greater depths to allow 100 thousand to 150 thousand and higher DWT supercolliers to call at U.S. coal ports would be a major step toward achieving such economies of scale for U.S. coal in export markets.

DREDGING COSTS

Port and harbor improvements are in themselves very expensive. The aforementioned ICE report states that to dredge three ports-- Hampton Roads, Mobile, and New Orleans--to 55 feet would cost in excess of $1.5 billion. The same report mentions that costs of improvements in New York would be approximately $140 million and in Baltimore $278 million. Other ports are equally as deserving of monies for port and harbor improvements. Obviously, the traditional methods of full financing of these improvements through general treasury funds are inadequate in view of the total amounts needed in a relatively short amount of time.

N.C.A. FUNDING PLAN

In recognition of this, and in recognition of the importance of making timely improvements, the National Coal Association has adopted the position that these improvements should be funded, at least in part, by port fees equitably applied, on a tonnage basis, on all commercial traffic in a port. Briefly the National Coal Association proposals which:

1. authorize dredging of ports and harbors to depths determined by the Corps of Engineers to be economical and practical; and
2. authorize individual ports to levy a port fee to fund, at least in part, the costs of dredging the port to those authorized depths.

These port fees should be characterized by the following:

1. they should be levied, on a port-by-port basis, according to broad guidelines set by the Congress designed to give an individual port some flexibility to meet the requirements of its own traffic patterns but designed to ensure that those fees are fair and equitable to all port users; and

2. they should be levied on a tonnage basis on all commercial traffic in the port whether for export, import or for intercoastal traffic.

But, if non-federal interests are to pay, at least in part, for these improvements, the non-federal interest should be assured that those improvements will be accomplished within a reasonable time-frame. Accordingly, any proposal calling for non-federal financing of port and harbor improvements should include a mechanism for expediting permitting procedures and the environmental review process for both groundside facilities and for the actual dredging work.

Fast-track permitting and expedited environmental review are activities which properly may be made the lead responsibilities of the Corps of Engineers in its traditional role with respect to port development projects. In regard to expedited environmental reviews, it is considered essential that a lead organization be designated to coordinate local, state, multi-state, and federal agency participation in the process. Both permitting and environmental review should be mandated for completion within a strict, relatively short time-frame, on the order of 180 to 270 days. The National Coal Association has detailed suggestions for expedited environmental review and will submit them for the record.

THE OVERSEAS MARKETS FOR
HIGH SULFUR COAL
GEORGE ECKLUND

The environmental and operating disadvantages of high sulfur
coal do not need repetition here. Generally, high sulfur coal is
unattractive to most users and is only taken under special
circumstances: 1) where the Btu value is high and sulfur is within
acceptable limits, or 2) where the sulfur can be absorbed in the
consumption process such as the cement industry, or 3) where the
coal can be obtained at a discount for blending with low sulfur
coal.

THE FUTURE FOR HIGH SULFUR COAL

The domestic markets for Illinois coal are being developed
through technology to reduce the impact of the high sulfur content.
This action is necessary to make full use of a vital regional
natural resource. Foreign customers, however, have full freedom of
choice in making their decisions, and they will buy low sulfur coal
as long as it is available at a reasonable price. High sulfur coal,
therefore, will find an overseas market only when the price spread
between low and high sulfur coal overcomes the disadvantages of
using high sulfur coal. This spread currently is not sufficient to
attract overseas customers. Increasing demand for the low sulfur
coals should cause their prices to rise, and this could force users
to reassess their use of high sulfur coal.
Foreign demand for Illinois Basin coal does exist, but it is
limited at present and markets must be sought assiduously. Because
of restraints of the Panama Canal, the Far East market is quite
restricted. The primary overseas market is in Western Europe.
In preparation for this conference, Zinder-Neris, Inc., polled
some of its representatives and friends in Europe and asked the
following questions:
1.   What is the absolute limit of sulfur which your utilities and
     industry will use?
2.   Are they prepared to buy some cheaper high sulfur coal to blend
     with low sulfur coals exported from elsewhere?
3.   What kind of price discounts would be needed to encourage this?

4. What users, such as cement plants or industry, do you have
    which can take higher sulfur coals? What percentage (rough
    estimate only) of your total imports would this represent? Do
    you expect this figure to increase or decrease in the future?
5. What are your views on the development and growth of scrubbers
    or fluid bed combustion units or other methods of burning high
    sulfur coal cleanly?
    The replies to these questions are summarized in Table 1.

DESULFURIZATION RESEARCH IN THE U.S.

Work is going forward in the U.S. on flue gas desulfurization
and fluidized bed combustion in order to develop a domestic market
for high sulfur coal. The Europeans are doing similar work, but
their primary objective is to control emissions rather than to
encourage use of high sulfur coal. Nevertheless, as a result of
their work, they may be able to take more high sulfur coal in the
future.

SLURRY AS A SOLUTION TO HIGH SULFUR

Another market is developing for high sulfur coal, but it is
not clearly defined at present. This involves processing a fuel
that would replace the heavy residual oil used in the Texaco and
other gasifiers. The gasifiers currently use a high sulfur resid,
and in the gasification process the sulfur is removed. Work is now
underway in Germany and Sweden that may result in replacing the oil
with a stabilized coal/water slurry. If this process succeeds, high
sulfur coal could be exported from the United States in slurry form,
representing a high Btu fuel at a reasonable price with no sulfur
restriction.

Table 1

PROSPECTS FOR USE OF HIGH SULFUR COAL IN WESTERN EUROPE[1]

Question	Denmark	Italy	Sweden
1) What is the acceptable limit of sulfur for utilities?	2.00%	1.00%	0.80%
2) Will utilities buy cheaper coal for blending?	Yes, but not all have mixing capabilities	To a limited extent	NA
3) What discount is required on high sulfur coal for blending by utilities?	100%	2% for each 0.1% of sulfur over 1.00%.	NA
4) What industries will take high sulfur coal? How much?	Cement: 100,000 tons annually of 2.5% sulfur	Cement: 1 million 1980. 2.5 m. 1985. 5 m. 1990.	Cement: 200,000 tons now, 300,000 tons 1985. Pulp: 300,000 tons max.
5) What are prospects for scrubbers, fluid bed units, etc.?	Limited	Studies underway, results unknown	Will be used but only for coal with 2% sulfur

[1] Responses by Zinder-Neris, Inc., correspondents in Western Europe, May, 1981.

(Table 1 continues on next page.)

Table 1 (Cont.)
PROSPECTS FOR USE OF HIGH SULFUR COAL IN WESTERN EUROPE[1]

Question	Germany	Spain	France	Belgium
1)	1.0% utilities 1.5% to 2% for industry	1.5%	1.0% 2.0% if very high Btu	2.0%
2)	Almost unknown in Germany	Many will if facilities available	Yes, if volatility roughly equal	NA
3)	Price discount immaterial in view of anti-pollution laws	Sufficient to offset extra cost	Must cover all costs of blending	NA
4)	Only cement, but established discount $4 per ton required	Cement: 30% of steam requirement	Cement: 2.3% to 2.5% sulfur	NA
5)	Technical improvements may encourage future demand, but in long term high sulfur coal will not exceed 5% of total coal use	Utilities will avoid expense of scrubbers and other high cost facilities	Utilities are not using scrubbers and have no plans for their use	NA

[1] Responses by Zinder-Neris, Inc., correspondents in Western Europe, May, 1981.

FORECASTS FOR THERMAL COAL
DEMAND IN EUROPE
ZACHARIAH ALLEN

It seems that the Midwest, with approximately 45 billion tons
of coal reserves having a sulfur content greater than 1.9%, has been
dealt a "double whammy" in recent years. First, our own environmental laws favored low sulfur coals either from the Powder River
Basin or the Appalachian areas in the East. While some of that
imbalance has been corrected, the impact of the correction has yet
to be felt in a major way.

More recently, we have seen dramatic growth in the need for
other major industrial countries, such as Japan and some of the
countries of Western Europe, to import thermal coal not only for the
generation of electricity but also for other industrial uses.
However, there is a general preference in these markets for coal
with sulfur content of less than 1.2%. This is due to the fact
that, with the exception of Japan, most of the other countries have
pursued a compliance coal concept in controlling sulfur dioxide
emissions from electric utility generating plants. This situation
is expected to change in West Germany and the Netherlands, where
some sort of desulfurization, probably scrubbers, will be required
in the future. However, as in the United States, that only applies
to plants that may be built in the future and, therefore, does not
represent an immediate market for our higher sulfur coal producers
today.

PREDICTED HIGH SULFUR COAL DEMAND

F. R. Schwab & Associates performed a market study of thermal
coal demand in Western Europe for the Interagency Coal Export Task
Force that issued its report earlier this year. Included here are
three tables of data (Tables 1, 2 and 3) that summarize the findings
of our report to the U.S. Department of Energy. Each table covers
our forecast for a particular time period, i.e., 1985, 1990, and
2000. Each forecast covers the estimated import requirements of 14
Western European countries and our estimate of where we think this
coal will come from. These tables are annotated to show that demand
is primarily for coal with a maximum sulfur content of 1.2%, unless

Table 1
FORECAST OF SOURCES OF SUPPLY OF THERMAL COAL[1]
TO MEET ESTIMATED IMPORT DEMAND IN WESTERN EUROPE (1985)
(MTCE)

Importers	Inter-EEC	Poland	Republic of South Africa	Australia
Norway[4]	-	0 -0.3	-	0 -0.3
Sweden	-	1.0-2.0	-	0.2-0.3
Finland	-	2.0-4.0	-	0 -2.0
Denmark	0 -0.6	0 -2.5	0 -3.5	0 -1.0
West Germany[4]	0.5-1.0	2.0-3.5	1.5-3.0	1.5-2.5
Austria	-	1.0-1.5	0 -1.0	0 -1.0
Belgium	0.2-0.6	0.5-1.2	2.0-3.5	0.6-2.5
Netherlands	0.2-0.6	0.5-1.2	0 -1.0	1.5-2.5
United Kingdom[2]	0 -0.5	0.5-1.0	0 -0.3	0.3-2.0
Ireland	0.3-0.8	0 -1.0	4.0-6.5	1.5-3.0
France	0.3-0.6	0 -1.0	4.0-6.5	1.5-3.0
Italy	-	2.0-3.0	3.2-4.0	2.7-4.0
Spain	-	0.8-1.4	3.0-3.2	2.0-3.4
Greece	0 -0.2	0 -0.2	0.2-1.0	0 -0.4
Totals[6]	1.5-4.9	10.3-23.3	13.9-27.6	10.3-25.5

[1] All coal demand is for 1.2 percent sulfur maximum except as noted.
   *1.2 to 1.8 percent sulfur for one-third of demand.
   **1.8 to 2.5 percent sulfur for one-third of demand.
[2] Possible exports from the United Kingdom to other Western European
   countries have not been included in this forecast.
[3] Includes Canada, Colombia, United States, and Inter-EEC.

(Table 1 continued on next page.)

Table 1 (Cont.)
FORECAST OF SOURCES OF SUPPLY OF THERMAL COAL[1]
TO MEET ESTIMATED IMPORT DEMAND IN WESTERN EUROPE (1985)
(MTCE)

Importers	Canada	Colombia	Other	United States	Estimated Imports**
Norway[4]	0 -0.3	-	0 -0.2[3]	0.1-0.3	0.65
Sweden[4]	-	-	1.0-2.0	-	3.3
Finland	0 -0.3	0 -1.0	0 -1.0	0 -3.0	4.0-7.0
Denmark	0 -0.4	0 -1.5	0 -0.7	1.0-2.0	10.7-11.1
West Germany	0.5-1.5	0 -1.0	0.5-1.0	1.5-4.0	10.0-10.5
Austria[4]	-	-	0 -1.0	0 -1.0	1.5-3.0
Belgium	0.1-1.0	-	0.2-1.0	1.5-2.5	8.3*
Netherlands	0.1-0.8	-	0.2-1.0	0.5-1.8	5.5
United Kingdom[2]	0.3-1.8	-	0.2-2.0	1.0-2.0	7.0*
Ireland	0 -0.6	-	0 -0.2	0.2-0.7	2.1*
France	0.5-2.0	-	1.5-2.5	2.0-3.0	12.7*
Italy	0 -1.1	-	0.3-2.0	3.2-4.0	14.6
Spain	0.3-0.6	-	0.4-0.6	3.0-4.0	11.0**
Greece	0 -0.4	-	0.1-0.5	0.2-0.8	1.4*
Total[6]	1.8-13.5	0 -3.50	4.4-15.7	14.2-29.1	92.75-93.15

[4] Swedish and Austrian figures are subject to change.
[5] Totals are taken from estimates of import demand for each country. The total for each country has been broken down into possible ranges for various sources of supply. In constructing this forecast, the minimal cannot exceed the total for each country.
[6] Totals for each country may come close to or exceed the tonnage FRS&A estimates will be available for exports. In such cases, the exporting countries may allocate the available quantities among their customers.

Table 2
FORECAST OF SOURCES OF SUPPLY OF THERMAL COAL[1]
TO MEET ESTIMATED IMPORT DEMAND IN WESTERN EUROPE (1990)
(MTCE)

Importers	Inter-EEC	Poland	Republic of South Africa	Australia
Norway[4]	-	0 -0.2	0 -0.2	0 -0.4
Sweden	0 -1.5[3]	0 -0.2	-	3.0-5.0
Finland	-	2.0-5.0	-	1.0-3.0
Denmark	0 -1.0	0 -4.0	3.0-6.0	1.0-4.0
West Germany	0.5-1.5	2.0-5.0	3.0-6.0	2.5-4.0
Austria[4]	-	1.0-2.0	1.0-3.0	1.0-3.0
Belgium	0.2-2.5	0.5-1.2	2.0-4.5	2.0-4.0
Netherlands	0 -1.0	0.5-1.4	0 -1.0	2.5-5.0
United Kingdom[2]	0 -0.5	0.5-1.5	0 -1.0	0.5-3.0
Ireland	0.3-2.5	0 -1.0	0 -1.0	0 -1.5
France	0 -2.0	0 -1.0	3.0-5.0	2.0-4.0
Italy	-	2.0-4.0	4.8-6.0	10.0-14.0
Spain	-	1.7-3.2	2.0-3.0	2.5-3.5
Greece	0 -1.0	0 -0.2	0.6-1.0	0 -0.6
Totals[6]	1.0-13.5	10.2-31.7	19.4-37.7	28.0-55.0

[1] All coal demand is for 1.2 percent sulfur maximum except as noted.
*1.2 to 1.8 percent sulfur for one-third of demand.
**1.8 to 2.5 percent sulfur for one-third of demand.
[2] Possible exports from the United Kingdom to other Western European countries have not been included in this forecast.
[3] From EEC sources.

(Table 2 continued on next page.)

Table 2 (Cont.)
FORECAST OF SOURCES OF SUPPLY OF THERMAL COAL[1]
TO MEET ESTIMATED IMPORT DEMAND IN WESTERN EUROPE (1990)
(MTCE)

Importers	Canada	Colombia	Other	United States	Estimated Imports[5]
Norway[4]	0 -4.0	0 -0.3	0 -0.3	0.1-0.3	0.75
Sweden[4]	0 -2.0	0 -1.0	1.0-3.0	0 -3.0	9.7
Finland	1.0-3.0	0 -1.0	0 -1.0	1.0-3.0	6.0-10.0
Denmark	2.0-3.0	2.0-3.0	-	1.0-4.0	13.7-14.4
West Germany	0.5-2.5	0 -1.0	1.0-3.0	2.0-6.0	20.0
Austria[4]	-	-	0 -1.0	0 -3.0	4.0-0.0
Belgium	0.5-2.0	0.5-1.5	0.2-2.0	2.2-4.5	15.3*
Netherlands	0.5-2.5	1.0-3.0	0.2-2.0	0.5-2.5	12.5
United Kingdom[2]	0.5-3.0	0.5-1.0	0.3-2.0	1.5-2.5	7.5*
Ireland	0 -1.5	0 -0.6	0 -0.5	0.4-1.5	4.5*
France	0.5-2.5	0.5-1.5	1.0-3.0	1.8-4.0	12.0*
Italy	0.8-2.0	2.0-3.0	1.0-3.0	4.0-8.0	37.7
Spain	0.5-1.0	1.0-2.0	0.5-1.0	3.0-5.0	13.8**
Greece	0 -0.8	0 -0.5	0.1-1.0	0.2-1.2	3.0*
Total[6]	6.8-26.2	7.5-19.4	6.3-23.8	17.7-48.5	160.45-170.15

[4] Swedish and Austrian figures are subject to change.
[5] Totals are taken from estimates of import demand for each country.
The total for each country has been broken down into possible ranges
for various sources of supply. In constructing this forecast, the
minimal cannot exceed the total for each country.
[6] Totals for each country may come close to or exceed the tonnage
FRS&A estimates will be available for exports. In such cases,
the exporting countries may allocate the available quantities
among their customers.

Table 3

FORECAST OF SOURCES OF SUPPLY OF THERMAL COAL[1]

TO MEET ESTIMATED IMPORT DEMAND IN WESTERN EUROPE (2000)

(MTCE)

Importers	Inter-EEC	Poland	Republic of South Africa	Australia
Norway[4]	-	0 -0.6	0 -0.8	0 -1.0
Sweden[4]	0 -0.2[3]	1.0-4.0	-	4.5-8.0
Finland	-	2.0-6.0	-	2.0-5.0
Denmark	0 -2.0	0 -5.0	3.0-6.0	2.0-5.0
West Germany[4]	1.0-5.0	2.0-6.0	5.0-11.0	5.0-11.0
Austria[4]	NO INFORMATION AVAILABLE AT THIS TIME			
Belgium	0.2-4.5	0.5-2.0	1.0-7.0	2.0-5.5
Netherlands	0 -0.3	0.5-2.0	0 -3.0	3.5-8.0
United Kingdom[2]	0 -0.5	0.5-1.5	0 -2.0	0.5-3.0
Ireland	0.3-3.5	0 -1.0	0 -1.0	0 -2.0
France	0 -3.0	0 -1.0	1.0-5.0	1.0-5.0
Italy	-	2.0-4.0	5.0-6.0	12.0-16.0
Spain	-	2.0-3.5	4.5-10.0	4.0-8.0
Greece	0 -1.0	0 -0.2	0.6-1.0	0 -1.4
Totals[6]	1.5-21.8	10.5-36.8	20.1-52.8	36.5-76.9

[1] All coal demand is for 1.2 percent sulfur maximum except as noted.
 *1.2 to 1.8 percent sulfur for one-third of demand.
 **1.8 to 2.5 percent sulfur for one-third of demand.
[2] Possible exports from the United Kingdom to other Western European countries have not been included in this forecast.
[3] From EEC sources.

(Table 3 continued on next page.)

Table 3 (Cont.)
FORECAST OF SOURCES OF SUPPLY OF THERMAL COAL[1]
TO MEET ESTIMATED IMPORT DEMAND IN WESTERN EUROPE (2000)
(MTCE)

Importers	Canada	Colombia	Other	United States	Estimated Imports[5]
Norway[4]	0 -1.0	0 -1.0	0 -1.0	0.4-1.2	2.7
Sweden[4]	0 -2.0	0 -2.0	1.0-5.0	0 -7.0	23.0
Finland	2.0-5.0	0 -1.0	2.0-3.0	2.0-5.0	10.0-18.0
Denmark	2.0-5.0	3.0-5.0	-	1.0-2.0	9.4-20.9
West Germany	1.0-10.0	0 -5.0	1.0-5.0	5.0-10.0	35.0
Austria[4]	NO INFORMATION AVAILABLE AT THIS TIME				
Belgium	0.5-5.0	0.5-3.0	0.2-6.0	4.0-8.0	24.6
Netherlands	2.0-6.0	1.0-4.0	0.5-6.0	2.5-10.0	34.8*
United Kingdom[2]	0.5-3.0	0.5-1.0	0.3-2.5	1.5-2.5	8.5*
Ireland	0 -2.0	0 -1.0	0 -0.5	0.4-2.0	6.5*
France	0.5-4.0	0.5-3.0	2.0-10.0	3.5-10.0	22.0*
Italy	3.0-7.0	4.0-6.0	2.0-6.0	7.0-12.0	45.0-66.0
Spain	2.0-5.0	3.0-6.0	3.0-5.0	8.0-15.0	33.0-50.0*
Greece	0 -1.4	0 -1.2	0 -1.2	0.2-1.4	3.6*
Total[6]	13.5-56.4	12.5-39.2	12.0-51.2	35.5-86.1	258.1-315.6

[4]Swedish and Austrian figures are subject to change.
[5]Totals are taken from estimates of import demand for each country.
The total for each country has been broken down into possible
ranges for various sources of supply. In constructing this forecast, the minimal cannot exceed the total for each country.
[6]Totals for each country may come close to or exceed the tonnage
FRS&A estimates will be available for exports. In such cases,
the exporting countries may allocate the available quantities
among their customers.

otherwise noted. Where noted, there is a tolerance for higher sulfur, up to 2.5%.

Looking at the table of 1985, one can see that the requirements for coal in Belgium, the United Kingdom, Ireland, France, Spain, and Greece allow for coal with over 1.2% sulfur. Out of a total esti- mated import demand of around 95 million metric tons (of coal equiv- alent), about 3.5 million tons (for Spain) will tolerate up to 2.5% sulfur in our judgment. An additional 14 million tons of demand will tolerate a range of 1.2 to 1.8% sulfur. In other words, about 18% of the forecast demand will be open to coals with sulfur in excess of 1.2% and only about 4% of the forecast demand will to- lerate coal with sulfur of 1.8 to 2.5%. We estimate that this percentage tolerating sulfur content over 1.2% will drop to about 14% of total estimated demand by 1990 and remain at that level through the year 2000.

The principal market tolerating higher sulfur levels is the cement industry. Since the limestone used in making cement can absorb the sulfur in the coal, up to a degree, cement kilns can use coal with sulfur content up to about 2.5 to 3.0%, depending on a number of process variables. However, the growth prospects for the cement industry in Europe are modest; hence, coal for that market will comprise a decreasing share of the thermal coal market there.

EFFECTS OF ACID RAIN

We have all heard about acid rain. The question of what causes it and what to do about it is getting wider and wider attention not only here in the United States, but also in Western Europe. Some countries, such as England and Germany. have pursued a policy of building tall stacks in excess of 200 meters in height to disperse SO_2. However, some of the downwind countries have begun to complain and these countries are demanding that this air quality control mechanism be reviewed. This re-examination may cause the development of more effective desulfurization technology in the future.

DESULFURIZATION TECHNOLOGY

A very real need exists to develop technology to remove sulfur prior to combustion. I am aware of some activities in this area that seem to have promise. Examples include the possibility of low Btu gasification in combined cycle applications and processes for treating coal chemically to remove the sulfur prior to the actual use of the coal. I have no doubt that there are many technical developments emerging in this area; which ones will prove commercially viable and environmentally effective is anyone's guess at present. However, it is in the best interests of all of us to encourage development of any and all possibilities in this area. The results will certainly benefit the U.S. since a large portion of our coal reserves contains significant percentages of sulfur content. We simply cannot afford to write off a large portion of our coal reserves because of sulfur content.

Unfortunately, such a technical solution affords no immediate relief to our midwestern coal industry. In the meantime, I would urge coal producers who do have the resources to take advantage of the high sulfur niche in the Western European marketplace to do so.

A REMEDIABLE IMPEDIMENT TO EXPORT SALES
OF U.S. HIGHER SULFUR COALS
THOMAS TRUMPY

The United States has very large reserves of coal with sulfur levels over 1.8%, the normal limit for use in power plants without flue gas desulfurization. Because desulfurization equipment is not yet used in power plants in most countries, the cement market appears the most attractive export market for such coal. Such coal has for many years been used in the cement industry in the U.S.A.

A substantial part of this medium or high sulfur coal is found in the Illinois Basin, covering reserves in Illinois, Indiana, and Kentucky. Most of this coal is of good quality having a calorific value of well over 11,000 Btu per pound. The mines are generally small and usually surface operations. The coal could be exported either via the Great Lakes or the Mississippi.

HIGH SULFUR COAL AND THE CEMENT INDUSTRY

Much of the cement industry in Europe is converting from oil to coal at present. In particular the plants in Spain, Italy and Greece, as well as other countries, are located close to the ocean ports able to handle imported coal. The quantities of coal required in most cases (for vessels under 50,000 tons) correspond to port capacities and shipment economies from the U.S.

The cement process is not disadvantaged by a coal of medium (1.8 to 3.0%) or high (up to 6%) sulfur content. The raw material for cement is a limestone. When heated in the cement kiln, the limestone absorbs substantially all the sulfur in the burning coal. The sulfur does not harm the quality of the cement. Higher sulfur coals can cause corrosion in the preheaters, but U.S. cement companies using coal of 3% sulfur or more claim to have resolved these problems by regulated air feed and temperature.

Therefore, to the extent that there is a price differential in favor of more than 2% sulfur coals, which may be in excess supply in the U.S., such coal, used in the cement industry in the U.S., could also be sold for export for cement manufacture.

VOLATILE CONTENT OF HIGH SULFUR COAL

The exportable higher sulfur coal from the U.S., particularly from the Illinois Basin, has over 30% volatile matter (often 32 to 36%). This causes no impediment to use but it does pose handling risks.

Coal is ground to a very fine powder for injection into a power plant burner or cement kiln. This powder is very combustible and may be explosive. The higher the percentage of volatile matter, the greater the risk of a sudden flash fire. This risk is highest as the dust leaves the grinding mill, where it is warm-air dried, or when it is exposed to oxygen, or stored for an extended period.

HIGH VOLATILE HANDLING TECHNIQUES

In the case of U.S. electric plants, the risk of explosion is reduced to a very low level by the use of the following:

1. well-designed and maintained equipment to avoid sparks and leakage (the National Fire Protection Association [NFPA] codes are a model of safety regulation);
2. inert gases only in the powdered coal cycle;
3. explosion doors in the powdered fuel handling cycle to permit combustion to vent harmlessly without spreading the fire; and
4. direct firing of coal into the burner as it leaves the grinder so that there is no unnecessary transport or storage of the hot coal dust (the roller or bowl mills used adjust more easily to variations in the firing feed rate and have far less coal in them exposed to the risk of blowback fires than do the ball-in-drum mills common in Europe).

EUROPEAN COAL HANDLING TECHNIQUES

Unfortunately, such handling techniques are not yet current in Europe, for Europe traditionally has used its own coal, and because this coal has only from 18 to 26% volatile matter, the explosion risk of this coal is lower than for Illinois Basin coals. The engineering firms which have been responsible for much of the equipment used to grind coal have therefore designed equipment on a traditional basis for these lower volatile coals (particularly Polish coal).

1. large rotating drum "ball mills" are used because they are cheaper to build, although they are also more energy intensive and carry larger charges of coal;
2. large storage bins ("day bins") are used to store powdered coal, permitting use of a large ball mill operating only part of the time (variation of throughput in such mills is very limited); and
3. hot powdered coal is carried from the grinder to the day bin by the mill air system with cyclone separator and fans.

While this system, exemplified by the engineering of F. L. Smidt of Copenhagen, has an acceptable measure of safety in damp northern Europe with low volatile north European coals, it is "explosively" unadapted to the new and growing use of coal (particularly in the cement industries) in the Mediterranean countries of Spain, Italy, and Greece.

Such potential users of coal have been advised that the only margin of safety in these ball mill/storage bin systems in hotter climates comes from the choice of a lower volatile coal (24 to 26% volatiles preferred, 29% limit). And this, of course, excludes the U.S. suppliers of otherwise well-adapted coal.

Some U.S. suppliers of coal are trying to blend coals to meet the ceiling on volatile matter, but this does not eliminate the presence of high volatile particles which could "backfire" on both the suppliers and the reputation of the U.S. coal industry.

Is there a solution to this problem? We are advised there is, and that U.S. coal need not be banned from the important export market for the cement industry (over 5 million tons per year for Europe by 1985).

REENGINEERING EUROPEAN COAL BURNERS

The solutions lie in reengineering European cement plants to direct firing (which is only bypassing the storage bin and is not expensive), avoidance of any storage of dry pulverized coal, safe construction of conveyor tubes and storage bins when they are required (use of "explosion doors" and heavy steel), and the use of inert gases in the powdered coal cycle.

These modifications are of very small cost (under $100,000 for a large plant) compared to the cost and security benefits to coal users to be able to include another source, U.S. coal, in their purchasing.

In the course of a series of studies prepared with F. R. Schwab & Associates, experts on the coal industry, we determined that lower sulfur coal will be in heavy demand by the middle of this decade (largely as coal use increases before desulfurization-equipped electric power plants are commissioned abroad).

The U.S. has a substantial and efficient mining industry with large exportable reserves of medium and higher sulfur coal. This coal can have an important export market if the U.S. government, the coal industry, and equipment manufacturers encourage the foreign cement industry to install equipment able to use a broad range of fuels, including U.S. high volatile coal. (For information sources of this article, see Notes.)

THE POTENTIAL WORLD DEMAND
FOR HIGH SULFUR COAL
RAY LONG

It's always useful in any discussion of international trade to remember that net exports or imports are the relatively small balancing item between two large quantities of supply and demand at the national or regional level. This makes for difficulty in forecasting and in keeping abreast of rapidly changing opportunities. The U.S. coal exporting situation is a case in point. As late as 1979, a DOE report to Congress listed exports as not much more than 100 million tons by the end of the century. Now, largely as a result of the World Coal Study--itself a major effort in international collaborative research--a possible level of 350-400 MT is indicated. And analyses of potential suggest long-run figures even higher.

This late development of awareness may reflect the fact that countries inevitably tend to look at their external trading oppor- tunities from the inside out, as it were. I detect a feeling in the U.S. now that the potential for a rapid expansion of the coal mining industry is there, but some demonstration of commitment to greater coal use is required on the part of the major importing countries. The importing countries, on their side, are somewhat wary of long- term reliance on imported coal for their incremental energy needs until ensured availability has been demonstrated. There is a danger here of the opportunities falling between two stools--of not getting enough done to generate the level of world coal trade suggested by global energy requirements over the balance of this century.

IEA COAL RESEARCH

The work of International Energy Agency (IEA) Coal Research, and the particular section of it for which I work, provide some global analysis of these issues, and I would like to start by saying something about my organization's work which is relevant to this presentation. IEA Coal Research was set up in London in 1976 shortly after the International Energy Agency was established in Paris. Its purpose is to look specifically at ways in which a greater use of coal can be promoted to replace imports of oil and gas in the major industrialized countries over the medium to long

term. The organization now comprises four services: 1) a Technical
Information Service; 2) a World Coal Reserves and Resources Data
Bank; 3) a Mining Technology Clearing House; and 4) the Economic
Assessment Service (EAS). The EAS is currently funded by 12 mem-
ber-countries of the IEA, including, of course, the U.S.

The EAS has concentrated its main effort to date on analyzing
costs of new coal-using technologies in more efficient coal-burning
for power generation; in gasification; and in liquefaction. In
addition, the service has done work on the economics of coal use in
residential and industrial markets, and on the costs of
environmental control in coal use. Finally, there is a section--in
which I work--called "Supply, Transport, and Trade," which is
looking at the economics of the entire coal chain from mine
production, through the land and sea transport of coal, to
consumption and delivered prices in the market.

My job in this has two aspects. One aspect has been to develop
a framework of thinking about global energy demand and supply, cost
and price, which clearly have an impact on the more specific
analyses of coal economics which EAS is carrying out. This is being
pursued by developing a regionally disaggregated global energy model
called EDSAFF (Energy Demand/Supply Analysis and Forecasting
Framework). This work is still in its formative stage, and some of
the numbers I will be putting up later are still subject to
reconsideration as the development of the model continues.

The other aspect has been to look specifically at constraints
on internationally traded coal. Some forecasts have come out of my
work on global energy demand and supply which are at the higher end
of the range of forecasts in the World Coal Study referred to
earlier. I am now in the process of examining possible physical and
institutional constraints which could prevent these high levels of
trade from being attained. So far, this work has concentrated on
the cost and availability of coal by sulfur and Btu categories from
major exporting countries. This is particularly relevant to the
discussions at this conference, so I want to concentrate my remarks
on future levels of international trade; on the role of coals of
varying sulfur categories in this; and on the position of U.S. coal
suppliers with some reference to Illinois coal and midwestern coal
in general.

WORLD ENERGY AND COAL FLOWS

The key factors influencing the level of international coal
trade are the following:
1. economic growth and its geographic occurrence;

2. energy dependence, defined as the rate of increase in energy demand relative to economic growth;
3. relative costs and prices of energy by type; and
4. energy substitution.

Table 1 summarizes the conclusions I've reached reached about points 1, 2, and 4; information on 3 will follow. The main points are a rapid rate of economic growth in the 1990s after the subdued growth of the 1970s continue into the 1980s; some improvement in energy dependence compared with what it has been historically (about 1:1 up to the late 1970s; a rapid expansion of nuclear power; and sustained increases in hydropower, gas, and coal. A lower rate of increase in world output or greater energy dependence would detract from coal demand. On the other hand, perhaps the more likely shortfall of nuclear power below the high growth rates envisaged would increase the importance of coal as the marginal energy supplier.

International trade is determined precisely by the imbalance, at the country or regional level, in the energy demand and supply in question. Table 2 shows a relatively large net importing potential on the part of the developing countries and an absolutely large importing potential on the part of the developed countries. The production forecasts for Western countries are based on detailed analyses of reserves and potential. I have then assumed that communist countries could make up the balance of supply required to meet demand.

That last assumption raises a host of questions regarding the technical feasibility of exploiting remote reserves, capital and technological transfers between the non-communist and communist world, and East-West geopolitical relations in general. But the fact is that between the Soviet Union and China, there are large reserves (209 out of 663 billion tons of recoverable reserves in the world according to latest World Energy Conference data), and it makes no sense to talk about a high coal-demand scenario without the vast reserves of the communist countries being brought into the world trade picture in some way. The point to make here is that if communist countries do not become significant net exporters of energy, even greater pressure for expansion is put on western coal producers.

Table 3 shows coal trade by major exporters and importers. Points to note here on the importing side are the rapid rise of West European and Japanese imports, and also the relatively large increase in coal imports to Southeast Asia, particularly in the 1990s and, to a lesser degree, to South America. On the exporting side, the export potential of the four major non-communist exporting countries is particularly interesting when seen relative to their domestic demand (Table 4). Exports from Australia account for 48% of total production, and from Canada and South Africa for 35-40%.

Table 1
ECONOMIC ACTIVITY AND ENERGY DEMAND
(ANNUAL % INCREASES)

	1973-77	1977-90	1990-2000	1977-2000
World gross product	+ 2.6	+ 2.5	+ 4.8	+ 3.5
World primary energy demand	+ 2.5	+ 2.2	+ 3.8	+ 2.9
Primary energy sources:				
Coal	+ 2.9	+ 3.8	+ 4.5	+ 4.1
Oil	+ 1.7	+ 0.5	+ 0.7	+ 0.6
Gas	+ 2.1	+ 2.0	+ 5.8	+ 3.7
Nuclear	+26.6	+ 7.9	+12.2	+ 9.7
Hydro	+ 3.7	+ 5.8	+ 5.3	+ 5.6
Non-Commercial	---	- 0.5	- 4.9	- 2.4
Energy dependence	0.97	0.88	0.79	0.83

Table 2
REGIONAL COAL REQUIREMENTS AND PRODUCTION
(MILLION TONS OF COAL EQUIVALENT)

	1977		1990		2000	
	Req.	Prod.	Req.	Prod.	Req.	Prod.
Developed countries	1157	1125	2547	2515	4328	4158
Oil producers	10	8	15	5	20	9
Other LDC's	135	128	153	143	446	209
Communist countries	1476	1518	1812	1864	2213	2631
World	2778	2778	4527	4527	7007	7007

Table 3
REGIONAL COAL TRADE
(MILLIONS TONS OF COAL EQUIVALENT)

Exporters			Importers		
	1977	2000		1977	2000
Australia	39	195	West Europe	77	709
Canada	12	100	Japan	62	219
South Africa	11	100	S.E. Asia	3	225
U.S.	49	367	Latin America	6	46
Communist	73	457	Communist	34	39
Other	18	54	Other	21	35
Total	203	1273		203	1273

Table 4
COAL DEMAND FOR MAJOR EXPORTERS IN 2000
(MILLION TONS OF COAL EQUIVALENT)

	Domestic	Exports	Total
Australia	215	195	410
Canada	188	100	288
South Africa	159	100	259
U.S.	2264	367	2631

In the case of the U.S., however, the proportion of production accounted for by exports, while rising from its current level of 7-8%, is still only 14%.

Despite having reserves large enough to expand production well beyond the levels indicated, there are signs in Australia, Canada, and South Africa that limitations could be imposed on exports for reasons to do with long-term national interest, balanced regional development, and availability of manpower. The implication is that the U.S., with less pressure of competing demands between the export and domestic markets, would have greater flexibility in meeting export opportunities.

QUALITY, COSTS, AND REGIONAL POTENTIAL

This leads us to the question of importer preferences, quality, and costs. The fact is that currently, major importers are tending to favor non-U.S. coals, partly perhaps because they don't want to replace dependence on Middle East oil with a comparable dependence on U.S. coal, but more specifically because Australian, Canadian, and South African coals for export are entirely low sulfur and can be delivered at a price--in the case of Australia and South Africa, certainly--that is less than U.S. delivered prices even where long sea journeys are involved.

An indication of production relative to capacity and cost at the levels of demand indicated earlier is given in Table 5. These numbers summarize the regionally disaggregated work I've recently done on coal availability by quality for the countries mentioned. They are subject to two major qualifications: first, the capacity figures relate only to capacity as currently documented; long-run capacity in all four countries is certainly much greater. Second, the cost numbers relate to 1981 costs and do not take into account future real increases in capital and labor costs.

Given these qualifications, at the levels of demand indicated earlier, it seems unlikely that total demand for U.S. coal, domestic and exports, at the end of the century can be met entirely from low and medium bituminous coal production; some of the demand will have to be met from a mixture of subbituminous low sulfur coal (LSC) from the West and bituminous high sulfur coal (HSC) from the East and Midwest. The two extremes of maximum western LSC and maximum HSC are indicated at the bottom of Table 5. The point is that U.S. production costs, already at a disadvantage relative to other exporters, tend to increase with the greater use of HSC as the balancing supply compared with western LSC, though these numbers take no account of the different costs involved in getting coal from mine to domestic markets or exporting port.

Table 5

PRODUCTION CAPACITY AND COSTS OF MAJOR EXPORTERS

	Production in 2000 (MTCE)	Percentage of Identified Capacity	1981 Costs in $ Per Ton
Australia	410	34%	25
Canada	288	66%	25
South Africa	259	44%	17
U.S.	2631[a]	79%	39
	2631[b]	98%	49

[a] Maximum production of bituminous LSC and HSC with balance from western subbituminous LSC.
[b] Maximum production of bituminous LSC and MSC with balance from midwestern and eastern HSC.

THE LONG-TERM ACCEPTABILITY OF HSC

Such as it is, this scenario is not very promising for HSC from the Illinois Basin in the short-to-medium-term. U.S. coal is generally high-cost compared with other potential suppliers to the export market. This is not to say that the U.S. will be priced out of the market for low sulfur coal. Given a policy on the part of the major importers to diversify sources as much as possible, the international coal market should operate like any market conforming to theoretical demand/supply relationships. Prices will be set by the highest-cost producer whose output is required to clear the market. What this does mean, though, is that importers will prefer only high sulfur coal if there is a substantial difference in price which offsets the costs of desulfurizing it. This does not seem to be a possibility in the near-to-medium-term, given the amounts of exportable coal from Australia, South Africa, and Canada which are entirely low sulfur and relatively low cost, even at higher levels of demand than indicated earlier.

We need to look, therefore, at the possibility of high sulfur coal beginning to be in high demand over the medium-to-long-term with the commercialization of new technologies which solve the environmental problems associated with its use. These technologies include beneficiation and scrubbing, fluidized bed combustion, gasification, and liquefaction. At this point, it is appropriate to review EAS' own work. As I said earlier, a major part of EAS work has been on coal conversion technologies. Broad cost estimates have been derived for new technologies in power generation, gasification, and liquefaction, and the main thrust of recent work has been reviewing the detailed economics of different gasification processes and indirect liquefaction methods.

Another area of work which looks specifically at the problems of sulfur control has produced cost estimates for a range of desulfurization technologies in the pre-combustion, combustion, and post-combustion stages of coal utilization. Without going into the details of this work, I'll try to summarize some conclusions that are being reached against the background of the macro-energy analysis which I presented earlier. While medium sulfur coals may be amenable to the less costly benefication and FGD systems, the widespread use of high sulfur coal awaits the commercialization of large-scale FBC and gasification/combined cycle in power generation; substitute natural gas; and liquefaction. The time-scale of the first two may be within this decade, though there is some question about how widespread their implementation might be.

A greater demand for HSC may await the 1990s and the early years of the next century when, under the pressure of high energy requirements, conventional sources of oil may begin to run out.

Then, the very high costs of unconventional sources of naturally occurring liquids, including enhanced recovery from conventional sources, will provide a greater possibility of economic coal liquefaction. The point has been made by the International Institute of Applied Systems Analysis, in a recent study looking well into the twenty-first century, that within the next 50 years, coal could become too valuable as a source of liquids to be used in its conventional applications, including power generation which would then be increasingly accounted for by nuclear, solar, and possibly other renewable technologies.

ECONOMIC GROWTH ASSUMPTIONS

The question that scenario gives rise to is whether the implied high costs of energy could be absorbed into the income-generating processes of economic growth without undermining growth itself. This brings us back to the central importance of economic growth assumptions in any forecasting exercise. One reaction to the forecasts presented earlier is to ask why they are more likely than others which are less bullish about economic growth and energy demand in the long term.

The answer is that it is oversimple to look on forecasting just as an exercise in the positive economics of predicting what will happen in the future. Forecasts need to contain a large element of normative economics because they are an expensive luxury unless policy formulation or management decisions are influenced somewhere along the line. Forecasting, therefore, has to be a three-stage process of identifying what could happen; what ought to happen to meet defined political or profit-oriented objectives; and what has to change to make the goal become the reality.

The high growth and associated energy demand forecasts reflect a world economy moving successfully, if slowly, out of the doldrums of the past few years toward recovery in the second half of the 1980s and renewed vigor in the 1990s--something applauded by all, presumably apart from the champions of nil growth. The forecasts may look too optimistic, but they are attainable if governments and businesses have the will to make them so. This is a principle, I suggest, which applies not only to the broad macro-energy developments which I have been describing, but also to the more local concerns about the future for Illinois coal. The market may not look good in the short term, but in the long term, energy--even coal--may be too scarce a commodity for consumers to be choosy about its qualities. (See Notes for additional information concerning this article.)

MARKETING MECHANISMS FOR THE
EXPORT OF HIGH SULFUR COAL
ARTHUR F. NICHOLSON

The need for additional markets for high sulfur coal does not
have to be documented anew. There has been ample evidence of
decreasing production and increasing unemployment in the Illinois
Basin coalfields since 1977. The rate of growth of coal-fired
electric generating capacity east of the Mississippi River has been
and will be too low to provide the needed markets. Other current
domestic markets for high sulfur coal are too small to be of any
consequence. On the plus side, however, the high cost of oil is
causing many industrially developed countries to turn to coal as a
fuel for their electric utilities.

The most prominent areas for export markets are the Pacific Rim
countries, i.e., Japan, Korea, Taiwan, Hong Kong, and Singapore and
Western Europe. Japan, with few domestic coal resources, has been
active in the development of technology to control the emissions
from burning coal, and along with the other Pacific Rim countries,
will be a growth market for coal. Unfortunately, the coal resources
of Australia, the western U.S., and western Canada are more
favorably located to serve the Japanese markets than is the Illinois
Basin's high sulfur coal. Because of transportation networks and
costs, Western Europe is a more logical market for Illinois Basin
coal, and the emphasis in this paper will be concerned with mechanisms to serve that market.

It could be argued, and probably will be, that no new marketing
mechanisms are needed for high sulfur coal. There are several coal-
producing companies and coal brokers who are heavily involved in
marketing coal to Western Europe. It is not our purpose to
criticize them for lack of knowledge or lack of effort to market
high sulfur coal. While there are some notable successes, clearly
there are not enough successes, therefore, some ideas that may be
useful in expanding the markets for high sulfur coal are suggested.

Except for limited specific situations, potential coal users in
Western Europe are looking for low sulfur coal. As described below,
the technology to burn high sulfur coal lags behind the U.S. experience. Being painfully aware of the slowness of U.S. electric
utilities to install equipment that would permit the utilization of
high sulfur coal, it is our feeling that significant growth in high
sulfur coal markets in Western Europe will develop slowly without

external stimulus. The external stimulus cannot simply be more sales effort. A different kind of sales effort is needed, one that is long term and that meets the competition in terms of cost, reliability, and environmental concerns.

The focus of the producers and brokers is necessarily short term. The pressures to produce result in the immediate future make it difficult for them to expend significant effort to develop markets that will absorb coal only after 2 two or 3 years of marketing effort.

PRESENT WESTERN EUROPEAN SOURCES

The competitors for the Western European coal market are the domestic coal resources (which range from almost none in Italy to substantial in England and Germany), and South Africa, Poland, Australia, South America, and other miscellaneous sources. South Africa is well situated to serve a limited share of the market, but must conserve its resources for its own energy needs. The problems with relying on Poland are well known. Australia is more logically a supplier to Japan, and South America is a limited supplier. The U.S. is the best and most reliable source of steam coal for Western Europe. Appalachian coal is the preferred choice at the moment by several coal users in Europe, quite possibly because Appalachian coals are somewhat similar to the design coals used in the coal-fired plants.

The implication is that if high sulfur coal is to become widely used in Western Europe, a significant effort will be needed to cause the boilers to be designed to use Illinois Basin high sulfur coal. For that to happen the utility building the plant or the company installing industrial-sized boilers must be convinced that the cost and reliability of using high sulfur coal is both equal to or better than using other fuels.

MARKET PRICE AND RELIABILITY

The cost aspect of the equation in terms of market price can be determined readily by competent and respected consultants. The problem is, as it always was, to establish valid assumptions about the nature of future costs of the different fuels. Therefore, reliability of supply becomes a highly significant factor.

Reliability of supply has even more overtones in an export market than in a domestic market where producer or transportation work stoppages are the chief concerns. In addition to these two, there are at least two other important considerations in an export market. Political stability in the exporting country and the likelihood of obtaining export permits are a significant consideration in reliability. Another is the willingness or capability of the seller to fulfill the contract when better market opportunities or unforeseen problems arise. This last consideration is especially important when small companies are involved.

The effort in Kentucky through the Department of Energy has been to develop a synfuel market in western Kentucky to provide a market for our high sulfur coal and to work on an export market for eastern Kentucky coal. East Kentucky is characterized by a large number of independent producers who individually cannot service the export market. Our initial thought was to establish a producers' cooperative that would include enough coal resources to support a long-term contract. However, a cooperative by itself is insufficient. There is still the problem of reliability of supply. A cooperative may not be strong enough to ensure that all members meet their contractual obligations. It was necessary to establish an authority, a nonprofit public corporation, to perform certain functions in the Commonwealth's best interests, and to act as a conduit between our independent producers and the export market.

TECHNOLOGY FACTORS AND THE EXPORT MARKET

At this point in the development of the Kentucky export market, we have urged the West European countries to plan to use high sulfur coal and have offered technical assistance through our coal laboratory. There is some polite interest, but the emphasis is still on low sulfur coal. The task will be to demonstrate clearly the economic advantages of burning high sulfur coal. We are developing the staff capability to perform that task and they are honing their skills on coal versus gas and oil fuels for industrial-sized plants.

In the comparison of technology and cost factors of differing air pollution abatement options, it can be seen that we need to make more cost-effective the options for using high-sulfur coal in Western Europe. It is imperative that we demonstrate the environmental and economic feasibility of synthetic fuels, of fluidized bed combustion, of coal-oil mixtures, etc., if we are to convince foreign buyers, whether utilities or industry, that Illinois Basin coal is a viable alternative to their present sources. This, plus the lower cost, compared to other fuels, and the reliability of the source, will make Illinois Basin coal competitive in the export market.

WESTERN EUROPEAN CONTROL STRATEGIES
AND ABATEMENT OPTIONS

As noted above, most European countries utilize low sulfur coal as a primary means of meeting air pollution control regulations. However, there are several future abatement options which offer potential for opening the European markets to higher sulfur coal. A marketing strategy Illinois Basin coal must have is the increased demonstration of the technological feasibility and cost advantages of these options.

PRESENT POLLUTION CONTROL STRATEGIES

Most of the European countries have adopted air pollution regulations. Sulfur dioxide and total suspended particulates are the two principal pollutants from coal combustion regulated in industrial countries. Ambient air quality standards or guidelines for these have been established in many countries, as shown in Table 1. The standards are in some cases considerably more stringent than in the U.S. The extent to which European countries have succeeded in meeting their standards has not been documented, but it is generally felt that improvements have been made for sulfur dioxide[1] (see Notes).

The legislative controls on sulfur emissions vary widely, as shown in Table 2. Most have explicit requirements for emission limits or for the sulfur levels in the fuel. Most countries also regulate stack heights to limit ambient concentrations. No legislation has yet been enacted to control long-range transport; however, limiting emissions has an indirect beneficial effect.

FUTURE ABATEMENT OPTIONS

The development of alternative methods for sulfur removal will be essential to the use of high sulfur (3-5%) Illinois Basin coal in Europe. The steam coals used there now are uniformly low in sulfur; the 1975 average for 12 countries varied between .7 and 1.4% sulfur.[2]

Due to large price increases and the continuing uncertainties of supplies, further increase in the use of low sulfur fuel oil appears unlikely. But substantial quantities of low sulfur coal are known to exist in major coal-producing countries. What is not clear are the political, social, and economic constraints that will determine the availability and price of low sulfur coals in domestic

Table 1
AMBIENT AIR QUALITY STANDARDS FOR SO_2 AND
PARTICULATES IN INDUSTRIAL COUNTRIES*
(All units in $\mu g/m^3$)

Group or country	Sulfur Dioxide Annual	Daily	Other	Particulate Matter Annual	Daily	Other
Canada	60 [b]30	300 [b]150	900 [b]450	70 [b]60	120	
European Community	[o]80 [p]120	[q]350 [r]250	[s]130 [t]180			
Finland	180	250	[d]720	180	250	[d]720
France		250			[a]350	
Germany, Fed. Rep.			[f]140 [g]400 [dh]650			[fa]200 [fa]400 [hk]750
Italy		260	[i]260		300	[i]200
Japan		100	[c]60		100	[c]40
Norway		200	[i]400		120	
Spain	150	400	[j]256 [k]700	130	300	[i]200
Sweden		[e]300 [f]200	[c]60 [i]750		120	[c]40
United States	80	365	[him]1300	75 [b]60	260 [b]150	([m])

[a] As of 1978.
[b] Secondary standard.
[c] 6-mo average.
[d] ½ hr average.
[e] Maximum value.
[f] Long-term average.
[g] Short-term average.
[h] Special zones.
[i] 1-hr average.
[j] 1-mo average.
[k] 2-hr average.
[l] 3-hr average.
[m] Various State standards also may apply.
[n] Also, half this value for > 10μ
[o] PM>40.
[p] PM<40.
[q] PM<150.
[r] PM>150.
[s] PM<60.
[t] PM<60.

Source: E. S. Rubin, "Air Pollution Constraints on Increased Coal Use by Industry. An International Perspective," J. Air Pollution Control Association, vol. 31, pp. 349-360, 1981.

Table 2
LEGISLATIVE CONTROLS ON SULPHUROUS
EMISSIONS IN SOME ECE COUNTRIES

	Planning	Investment Decisions	"Polluter Pays"	Financial Incentives	Nuisance & Health Laws	"Alert" periods	Low S fuel, nationwide	Low S fuel, SPZ's	Low S fuel, "Alerts"	FGD Option	Emission Limits	Air Quality Criteria
AUSTRIA							?					X
CANADA											X	X
CSSR	?		?					X	X		?	
DENMARK	X			X			X	X				
FINLAND	X				X		?				?	?
FRANCE				X		X	X	X	X			
GFR	X					X	X			X	X	
HUNGARY	X	X	X					X			X	X
IRELAND	X						X					
ITALY							X				X	X
LUXEMBORG							X					
NETHERLANDS							X			X	X	
NORWAY	X		?				X	X		X		
POLAND		X	?								X	
SWEDEN							X	X				
SWITZERLAND							?					
USSR	X		X								X	X
UK	X			X	X		X	X			X	
USA											X	X
YUGOSLAVIA		X	X		X						X	

X Implemented wholly or partially
? Implementation under consideration

Source: R. A. Barnes, "The Long Range Transport of Air Pollution. A
Review of European Experience", Journal of the Air Pollution
Control Association 29(1979):1219-1235. Used with permission.

and international markets. A number of abatement strategies that offer an option to the use of low sulfur fuels are discussed below.

Stack Height

Tall stacks are an effective means of reducing local concentrations of sulfur dioxide. This means of control has been widely used in England and, to a lesser extent, in other countries, e.g., France and Germany. This is an inexpensive and technically simple control measure. As no reduction in total sulfur emissions is achieved, long-range transport problems will not show any concomitant improvement; in fact, they may be somewhat exacerbated by taller stacks.

Flue Gas Desulfurization (FGD)

FGD has received much attention in the United States, particularly in regard to large coal-fired electric generating plants. Industrial application in the U.S. has been growing slowly. None of the European countries has made the commitment to FGD technology. In fact, in 1978, only six full-sized FGD installations could be counted, three in Germany, and one each in Norway, Sweden, and France.[3] The technology is still viewed as too costly and technically unproven for widespread application.

Fluidized Bed Combustion (FBC)

FBC technology offers the promise of efficient sulfur removal at low cost. The process involves, however, some environmental tradeoffs, such as increased production of solid waste. More importantly, full-sized units are only now in the planning stage and it is unlikely that commercialization will occur before the 1990's. Partially offsetting this disadvantage is the fact that FBC may be cost-effective for retrofitting existing boilers.

Synthetic Fuels

Also on the horizon are technologies for the conversion of coal to clean gaseous, liquid, and solid fuels through the processes of gasification, liquefaction, and solvent refining. Each of these technologies is under development in the U.S. and one, the SASOL Fischer-Tropsch Indirect Liquefaction Process, is in commercial operation in South Africa. Small-scale gasification is already being practiced in Europe and elsewhere and may provide the earliest alternative fuel for industrial use. Each of these technologies is suitable for high sulfur Illinois Basin coals. Again, widespread application of synthetic fuel processes probably will not occur until after 1990 in the U.S. and even later in Europe.

Coal Cleaning

The mechanical cleaning of coal is best known as a means of reducing the ash content and increasing the heating value of
run-of-mine coal. With increasing concern over sulfur dioxide,
these processes are being reexamined for their sulfur removal
potential. The method is effective for pyritic but not organic
sulfur removal. Illinois Basin coals typically contain about 50%
pyritic sulfur. Thus, reductions to the equivalent of 2-3% sulfur
coal are possible, which is still high by European standards.
Washed coals could be blended with low sulfur coal to meet specific
sulfur limits. Although coal washing for sulfur removal is not
viewed as cost-effective in England, countries like Germany and
Poland have planned greatly expanded coal washing capacities.[4] In
the future, chemical coal cleaners using reagents like oxygen,
chlorine, or ferric iron may allow high removal of organic and
pyritic sulfur. Such processes do not appear to be near
commercialization yet. The cost of chemically cleaned coal is
expected to be high.

Coal-Oil Mixture

High sulfur coal may be mixed with oil to produce a low sulfur
fuel called "coal-oil mixture" (COM). COM can either be transported
to or mixed at or near the site of use. Boilers can be retrofitted
to burn COM, but existing oil or gas boilers will suffer some
derating. The process is technically simple, but many potential
problems (e.g., tube corrosion and particulate control) have yet to
be evaluated.

In summary, the most critical mechanism needed to market high
sulfur coal is to demonstrate that high sulfur coal can be burned in
accordance with environmental standards at a competitive cost and
that the coal will be there when needed. More can be done and must
be done to convince buyers from overseas markets.

ILLINOIS COAL EXPORT POTENTIAL
AND INTERNATIONAL PRICE COMPETITIVENESS
C. KENT GARNER

Regarding the supply/demand nexus, the fundamental theorem of
the steam coal market is that the market is demand driven. That is,
an increase in supply will not unilaterally cause increased
demand--rather, greater demand generates increases in supply.
Several recent, well-publicized studies of world coal demand, such
as The World Coal Study, the International Energy Agency County
Reviews, and the Interim Interagency Coal Export Task Force Report,
though differing in their specific demand projections, all concluded
a significant increase in world steam coal demand is virtually
certain. The demand over the next 20 years projected by the
Interagency Task Force for Western Europe and the Pacific Rim alone
rises from 88 million short tons in 1979 to between 475 and 565
million short tons in the year 2000. This demand growth equates to
overall annual compound growth rate of 8.8%-7.2% for Western Europe
and 13.1% for the Pacific Rim. In the year 2000, Western European
demand is expected to range from 273 to 343 million short tons while
projected Pacific Rim demand ranges from 202 to 222 million short
tons (see Notes).[1] It is reasonable to assume the factor essential to stimulate the international steam coal market--strong
demand--will evolve and, in fact, is materializing at this very
moment. Of course, the quality specifications required by utility
and industrial consumers will determine, to a large degree, the
extent Illinois producers participate in this market.

LOW VS. HIGH SULFUR COAL

According to the Interagency Task Force, foreign steam coal
consumers will require, in the near term, low sulfur (less than
1.5%) and high Btu (greater than 11,000 Btu per pound) steam coal
for their existing and planned boilers and kilns. These
specifications are required because, to date, outside North America
the primary methods of achieving compliance with emission standards
have been the limiting of fuel sulfur levels and the use of
electrostatic precipitators. Scrubbers have not been widely

utilized. More significant is that, with the exception of Japan, scrubbers are not presently included in the near-term plans for future utility and industrial power plants and kilns. Thus, European and Asian consumers will continue to rely on relatively low sulfur coal.[2]

Beyond the present planning horizon, in addition to relative fuel economics, national politics, and economic policies and growth will determine whether individual countries choose to bear the high front- end capital costs of scrubber installation and the resultant near-term effects on their balances of payments or opt for a more gradual economic adjustment process through paying higher prices for low sulfur coal, although this latter course may ultimately be more costly and may not minimize energy source dependence.

ILLINOIS COMPETITIVENESS

I would now like to review the general competitiveness of Illinois producers in Western Europe--the steam coal market which offers the best opportunities for Illinois coal. Over the first six months of 1981, the Coal Week quoted FOB mine marker price for Illinois spot sales ranged from $20-$22 per short ton for 10,500 Btu per pound, 3.5% sulfur, and 13.0% ash coal and $25-$27 for 11,700 Btu per pound, 2.5% sulfur, and 8.5% ash coal. For term purchases, the range was $21-$23 for the lower Btu, higher sulfur product and $27.50-$29.50 for the higher Btu, lower sulfur coal.[3] Recall that these prices are indicative of the state of the market rather than actual transactions, that they reflect labor wage rates under the 1978 UMWA contract, and that they include an implicit seller's profit margin. Taking the midpoint of the spot price ranges and applying a worst-case sulfur penalty of 50¢/ton per 0.1% sulfur content above 1.5% and making a proportional Btu adjustment based on an 11,500 Btu per-pound requirement, the following adjusted FOB mine prices result:

low Btu, higher sulfur--$10.04/ton
($21--(20 x .5)) x (105 : 115)
high Btu, lower sulfur--$21.37/ton
($26--(10 x .5)) x (117 : 115)

At today's production costs, these prices are not attractive--a problem I will return to shortly. But, how do these adjusted prices compare on an energy-equivalent basis to other coals landed in Europe?

ILLINOIS VS. OTHER COALS IN EUROPE

Transportation to a Gulf Coast port by rail or barge would add $12 or $13, respectively, per ton. The barge/freight figure includes charges for mine-to-dock truck freight of $3/ton (15¢/mi. x 20 mi.), a loading charge of $1.00/ton, and an imputed transfer loss charge of $0.50/ton. Adding loading and unloading port charges totalling $4.50/ton and ocean freight, excluding any demurrage charges, of $12-$14/ton, places spot Illinois coal on Western European piers for the following adjusted prices: lower Btu, higher sulfur for $38.50-$41.50/ton; and high Btu, lower sulfur for $50-$53/ton. Using the midpoints of these ranges, Illinois coal can be landed at a cost of $1.90/million Btu and $2.20/million Btu, respectively. By comparison, similar computations show South African and Australian coals representatively have a spot landed price of about $2.46 and $3.37, respectively, per million Btu.[4]

It seems clear that Australian steam coals are not price competitive in Western Europe and, assuming continued real price increases in ships' bunkers, they will likely remain uncompetitive due primarily to their sensitivity to ocean freight rates. Conversely, South African steam coals are competitive especially in view of their low sulfur content. And, South Africa's port throughput and ship loading capacities will insulate it from rises in bunker oil prices and be compatible with the increasing ship size capacities of major European coal ports. However, utility requirements for long-term, secure, consistent coal supplies evidenced by their reliance on long-term purchase contracts have already raised questions among some consumers of the dependability of South African suppliers in view of the country's political dilemma. And, Poland's reliability is of even greater concern. It is worth noting that consumers of other energy forms have always been willing to pay a premium for secure sources of supply.

As noted earlier, the adjusted FOB mine price for Illinois coal of $10 and $21 per ton would hardly cover current production costs let alone a profit for the producer. A recent study sponsored by the Electric Power Research Institute found that an FOB mine price of $23 to $33 per ton would be required to operate a new high sulfur Illinois coal mine profitably and that new mine development is being thwarted by current excess production capacity which is preventing producers from pricing coal in this range.[5] Given the demand-driven character of the market, price increases are unlikely until the current excess capacity is absorbed. Once it is, Illinois producers should be well positioned within the Western European price structure since a landed adjusted price equal to that of South African coal on an energy equivalent basis yields a quality-adjusted

FOB mine price in Illinois of $28 per ton.[6] Thus, in the absence
of excess U.S. capacity, the price presently being offered by South
African producers to Western European consumers would support new
mine development in Illinois and still permit some additional
quality discounting. On an energy equivalent basis then, Illinois
Basin coal is price competitive in Europe today!

POSITIVE FACTORS FOR ILLINOIS COAL

Several factors will affect Illinois' future competitiveness.
First, Illinois Basin producers are likely to be less sensitive to
inland freight rate changes than Appalachian producers--Illinois'
primary U.S. competitors in the Western European steam coal market.
A sensitivity analysis by ICF of Washington asserts 1990 U.S. demand
for midwestern coal would drop by only .5% in the event of a 33%
rise in transportation rates over that period while demand for
Appalachian coal would fall by 6.2%.[7] Though the ICF analysis
focused on domestic demand, the comparative insensitivity of demand
for Illinois coal to inland transportation costs would also show up
at export locations, though perhaps to a lesser degree. Second, any
freight rate increases which occur should be comparatively less than
those of other regions since Illinois Basin producers will doubtless
benefit from the competition--intermodal and intramodal--due to
being served by ten Class I railroads and numerous barge ports on
four major rivers.

Another factor which should increase the competitiveness of
Illinois coal is the productivity of Illinois' mines. The 1980
average mine productivity in Illinois was higher--both surface and
underground--than any Appalachian state's average except Kentucky's.
More important is the fact that 1980 Illinois surface and
underground mine productivity increased over 1979 by 29% and 19%,
respectively. Kentucky's surface mine productivity decreased 3%
while underground productivity fell by 7%.[8] Continuing
productivity gains will strengthen the competitive position of
Illinois coal in world and domestic markets.

A final factor which may emerge in the world steam coal market
to benefit Illinois producers is the inclusion of ash penalty
clauses in purchase contracts. Ash disposal constraints in densely
populated coal consuming nations, such as Japan and the Netherlands,
have already led to their inclusion in some recent contracts.
Illinois producers, on the average, should be less affected by this
than their U.S. competitors in view of Illinois coal's lower ash
content.

   In closing, despite hurdles yet to be overcome--current excess
U.S. production capacity, river and port loading and vessel
constraints, and the development and utilization of blending
technology to open wider markets--the future export potential for
Illinois coal is good, though not to be taken for granted. To
penetrate this growing market, Illinois producers must educate
European consumers about the advantages they can offer, which may
well offset market concerns regarding Illinois coal's sulfur
content.

IV

The Availability of U.S. High Sulfur Coal
to the World Energy Market: Industrial Perspective

REVIEW AND SUMMARY
WILLIAM W. MASON

COAL PRIOR TO WORLD WAR II

Prior to the onset of World War II, the U.S. had exported coal
from its eastern fields for nearly 100 years. While our export
market for coal has not been a major market-- averaging about 10% of
our annual production--it has been an important market. Prior to
and following World War I, coal was exported to Europe in fairly
good quantities for steam and heating use. Coal was exported all
through the 1930s in varying quantities, primarily to Europe and
Canada. During this period, too, most of the coal exported was
designated as steam coal, with the bulk of it moving through our
eastern and lake ports. In 1947, following World War II, we had a
banner year when we exported over 68 million tons. This was mostly
steam coal shipped from our eastern ports and mostly in 10,000-ton
liberty vessels. To a great extent, this coal moved to Europe for
the rebuilding and reconstruction of that war-torn area.

COAL DEMANDS FROM 1950-80

In the 1950s, 1960s and 1970s, with the exception of Canada,
the demand shifted from steam coal to coking coal. This was the
period of the development and construction of the many steel mills
around the world that used, and required as a base, coking coals
from the United States. Japan became the largest single receiver of
American coals, and in 1975 received from the United States over 27
million tons, or over 40% of all the coal exported. During this
period, coking coal was shipped to almost every country in Europe
with a steel plant and also to developing areas of South America--
Brazil, Argentina, and Chile.

SHIFT FROM OIL TO COAL IN 1979-80

It was not until 1979 that the market overseas began to shift
again to steam coal. In 1979 out of a total of 51 million tons

exported, excluding that to Canada, only about 2.4 million tons were categorized as steam coal, the remaining being coking or metallurgical coal. Why did this take place? If you will recall, this was the year of the big jump in oil prices. Europe, which was almost entirely dependent upon mideastern oil, concluded that the time had come when it could no longer depend upon, nor pay the price of, energy from oil and looked to alternate sources. Coal, and American coal in particular, appeared to be the answer, so in 1979 and 1980 we had many delegations visit our country to explore the possibilities and potentialities of securing coal for the conversions and expansions of their utility and steam requirements. Japan and the Far East, also heavily dependent upon foreign oil, began to look to other fuel sources; and while Australia, western Canada, Poland and South Africa were their more natural sources of supply, they too began looking to the U.S. as a source.

CURRENT STATUS OF U.S. EXPORT MARKET

In 1980 we exported almost 90 million tons of coal. Of this, 63 million tons were metallurgical, with the remainder steam, which was a great increase from the previous year. This year, 1981, we had earlier predicted and expected to ship about the same amount as last year; however, with the UMWA shutdown, this forecast will no doubt be downgraded, although I feel that with the resumption of operations, the export market will pick up rapidly. I must say that we have lost some credibility among our foreign customers and potential customers by our inability to perform in the manner in which they would desire during these last couple of months.

ISLAND CREEK'S MINING AND TRANSPORTATION

I would like also to tell you a few things my company is doing with regards to mining and transportation that I feel will assist in the marketing of Illinois Basin coals. At Island Creek we currently have recoverable reserves of 3.8 billion tons. We are bringing on new mines which will increase our production capacity beyond 33 million tons per year. Our plans call for continued growth, through the 1980s and ensuing years, in production and distribution terminals. The concept to develop river terminals and other transportation needs at Island Creek came about in 1977. Our long-range and strategic planning revealed that we needed to become less dependent on historical distribution methods that restricted our growth and affected our ability to mine and deliver coal.

RAILROAD AND RIVER PLANS

We therefore embarked on a program to define the transportation bottlenecks that were adverse to our marketing plans. To this extent, we first acquired 525 railroad cars, at a capital investment of 16 million dollars. Next, we turned to Mr. Reliable, "Old Man River." We found the barge lines most cooperative in our plans for increasing coal haulage. Our river coal shipments amount to several million tons annually. Our examination of the river system told us it had the capacity and could handle increased volumes. The rivers had demonstrated in the past the ability to handle some 130 million tons annually, despite certain interim lock and dam bottlenecks. We felt it was truly our best avenue for the expansion of markets. We also saw the need for adequate intermodal facilities.

RIVER TERMINAL DEVELOPMENT

We then began our development of new river terminals for the movement of additional coal to both the domestic and export markets. At present, there are three river terminals in operation. Their capital investment approximates over 70 million dollars. These are under the operation of Occidental's subsidiary, Kentucky-Ohio Transportation Company (KOT):

1. Big Sandy Terminal is located on the Big Sandy River south of Catlettsburg, Kentucky. It currently receives its coal by highway.
2. The South Shore Terminal, located at South Shore, Kentucky, on the Ohio River is served by the highway and the Chesapeake and Ohio Railroad. It was completed in April, 1980, loading out its first barges and rail coal.
3. The Wheelersburg, Ohio, Terminal is also located on the Ohio River. Served by the Norfolk & Western Railway, it began receiving coal in December, 1980.

These terminals, for the most part, load 15-barge tows hauling 22,500 tons each with sufficient fleeting areas for jumbo and even 3,000-ton super barges. The barges navigate the Ohio, Tennessee, and Mississippi River systems to domestic, utility and industrial plants. Coal will also be transshipped through the Gulf for export movement. In addition, KOT is developing a coal storage terminal at Baltimore in the area commonly known as Curtis Bay. Not only will KOT have a sophisticated transportation and ground storage facility, but also its own ship and barge pier to go with it. Construction has been underway since July, 1980. Currently under extensive study is development of an additional facility in the Portsmouth, Virginia, area.

As you can see, Occidental has backed its hand through investment in railcars and terminals for building a transportation network to support our coal marketing programs. There is still much to be accomplished in order to achieve a more efficient program for the movement of U.S. coals. This includes eastern and western, both steam and metallurgical.

FUTURE COAL DEMAND

Now let us move to the future and see what is needed. What will the future hold? According to the World Coal Study, the Interagency Coal Export Task Force, and National Coal Association forecasts, there is a healthy demand worldwide for steam coals. While these and other studies may disagree on an exact level for this demand, they do concur that the "order of magnitude" is sufficient to warrant an expansion in this country of our marketing capabilities directed toward meeting overseas demand. We already have sizable excess production capacity to support increased exports, and we can readily add to this capacity as needed.

To demonstrate this "order of magnitude," consider these ranges of tonnage forecasts starting with 1985:

1. For Western Europe alone, the imported steam coal demand is forecast in the range of 100 to 125 million net tons. By 1990, this range is expected to be 150-190 million tons. By 2000, the range is expected to be 270-340 million.
2. Even if imports actually fall on the low side of these ranges, they clearly demonstrate that a large market for steam coal is rapidly developing in Western Europe alone.
3. Turning to the Pacific Rim area, which includes Japan, Korea and Taiwan, projections also abound, but the ranges of these projections are not as extreme. For 1985, steam coal import requirements are forecast at 40 million tons; the 1990 forecast is 90 million, and by the year 2000, imported coal demand is projected to be above 200 million tons.

Totalling these projections and using the low figures in the ranges, we obtain the following forecast figures for imported steam coal demand in these two major overseas markets, Europe and the Pacific Rim:

1985---140 million tons
1990---240 million tons
2000---470 million tons

The National Coal Association report also makes this projection for world metallurgical and steam market import requirements. The forecast shows a total import range of 274 to 343 million tons of coal for 1985 as compared to 243 million tons in 1979. By 1990, this range is projected to increase from 382 to 467 million tons. The U.S. is expected to supply 30% of this demand. Other studies concur that the U.S. could supply one-third of the world's coal demand from imports.

INCREASING PORT FACILITIES, RAILROAD STOCK, AND PIPELINES

For the U.S. to increase exports over the long term, additional terminal and transportation capacity will be required. Terminals are being added at Baltimore. Others are being planned at Norfolk and at other East, Gulf and West Coast areas to be served by barge and by railroad. In addition, a number of improvements have been made at existing facilities of the Chessie and N & W at Hampton Roads, along with those in the New Orleans and Mobile area. New river terminals have also been developed to support export movements via our inland waterways. Therefore, affirmative action is being taken to increase our nation's future port capacity. At last count, some 37 terminals already have been announced as being built throughout the country with a capacity of 479 million tons by 1985. How many of these facilities will materialize and at what actual capacity remains to be seen (see Table 1).

By no means do we downplay what the railroads have done in improvements to their plant and rolling stock, nor the equipment investments being made by the trucking and barging companies. Pipelines are also being considered to transport the coal in slurry form. These elements contribute significantly and are essential to our delivery capabilities.

GOVERNMENT, PRIVATE INVESTMENT, AND DREDGING

While the private sectors have reacted expeditiously to the obvious market potentials and have begun to commit investment dollars, there still needs to be accelerated action on the part of government. Take dredging! Currently, all of our major coal ports, and all of those proposed, have channels of 45 feet deep or less. This means that the maximum size of vessel that can depart fully loaded is Panamex size on the order of 60,000 to 70,000 tons. Only one port, Hampton Roads, can handle vessel loads of 100,000 tons or just slightly greater. Ocean freight costs per ton are $5-$7 per

Table 1
EXISTING, PLANNED, AND ANNOUNCED PORT
CAPACITY FIGURES (IN MILLIONS OF TONS)

	1981/82	1983/84	1985/86
EAST COAST			
Port of Quebec, Canada			
1. Cast Group	6 MM	20 MM	20 MM
Albany, New York			
2. Atlantic Cement	1 MM	2 MM	2 MM
3. New Amsterdam Coal	2 MM	4 MM	4 MM
Fairless Hills, Pennsylvania			
4. United States Steel	-	3 MM	10 MM
Camden, New Jersey			
5. Alla-Ohio Valley Coal	2 MM	3 MM	3 MM
6. Camden Coal Terminal	2 MM	3 MM	4 MM
Philadelphia, Pennsylvania			
7. ConRail	4 MM	10 MM	10 MM
Baltimore, Maryland			
8. Chessie System	10 MM	12 MM	12 MM
9. Kentucky-Ohio Transportation	-	10 MM	10 MM
10. Consol	-	10 MM	10 MM
11. Pittston Co., Mapco, Inc., Elk River Resources, Old Ben Coal Co., Utah Int., Soros Assoc. (Anne Arundel County)	-	-	15 MM
Hampton Roads, Virginia			
12. Chessie System	14 MM	15 MM	16 MM
13. Norfolk & Western Railway	32 MM	35 MM	35 MM
14. Cox Property at Portmouth to be developed either by A.T. Massey, Pittston Co., Consol and Island Creek or the N&W Railway	-	10 MM	25 MM
15. Ashland Oil, Inc., Westmoreland Coal Co., Utah International, Inc., and Armco Inc.	N/A	N/A	N/A
Morehead City, North Carolina			
16. Alla-Ohio Valley Coal	3 MM	6 MM	15 MM
17. Gulf Interstate Coal Terminals	-	10 MM	20 MM

	1981/82	1983/84	1985/86
Wilmington, North Carolina			
18. Atlantic Resources, Inc.			
(American Coal Export Co.)	2 MM	4 MM	8 MM
19. Utah International	N/A	N/A	N/A
Charleston, South Carolina			
20. Massey Coal Terminal	2 MM	4 MM	8 MM
Savannah, Georgia			
21. Savannah Coal Terminal, Inc., Harbert, Elk River and Family Lines)	-	5 MM	15 MM
22. Hutchinson Island Coal Co. (Strachan Shipping Co. & Pope Evans Robbins, Inc.)	-	6 MM	12 MM
23. Southern Minerals Corporation	4 MM	8 MM	13 MM
Total East Coast: Eleven port areas with 24 coal handling facilities.	84 MM	192 MM	282 MM
GULF COAST			
Mobile, Alabama			
1. McDuffie Terminal	5 MM	10 MM	20 MM
New Orleans, Louisiana			
2. International Marine Terminals	4 MM	9 MM	12 MM
3. Electro Coal Transfer Corporation	10 MM	15 MM	20 MM
4. River & Gulf Transportation	-	12 MM	20 MM
5. Port of New Orleans/Ryan Walsh Stevedoring	2 MM	5 MM	10 MM
Baton Rouge, Louisiana			
6. United States Steel	-	12 MM	15 MM
7. Louisiana Coal Services	-	5 MM	10 MM
8. United Energy Resources	-	8 MM	15 MM
Steeltown, Texas			
9. Texas-Oklahoma Port Company	5 MM	5 MM*	5 MM
Galveston, Texas			
10. Galveston Port Authority	-	10 MM	20 MM
Total-Gulf Coast: Five port areas with ten coal handling facilities.	26 MM	91 MM	147 MM

*Note: With the addition of a second stacker-reclaimer, capacity can be increased to 10 million tons per year.

	1981/82	1983/84	1985/86
WEST COAST			
Long Beach, California			
1. Metropolitan Stevedore Company	2 MM	5 MM	15 MM
Los Angeles, California			
2. Port of Los Angeles-			
Terminal Island	3 MM	12 MM	20 MM
Port of Kalama, Washington			
3. Pacific Resources, Inc.	-	5 MM	15 MM
Total-West Coast: Three port areas			
with three coal handling facilities.	5 MM	22 MM	50 MM
Grand Total-Port Capacity			
for U.S. Coal	115 MM	305 MM	479 MM

ton lower in a 150,000 ton vessel as compared to those of 100,000 tons. Freight costs quickly escalate for vessels below 100,000 tons. Our world competitors now have, and continue to build, facilities to load these large colliers.

The Army Corps of Engineers has the responsibility of dredging and maintaining channels and harbors, but before work can begin deepening those channels, feasibility studies and environmental impact statements must be prepared and approved. Permits must be obtained. Finally, monies must be authorized and appropriated by the Congress for these river and port projects. We are encouraged by the number of proposals introduced in the 97th Congress thus far, which among other things would accomplish the following:

1. dredging harbors and channels on the East, Gulf, and West Coasts to depths of 55 feet or more; and
2. fast-tracking the environmental review and permitting process for both shoreside facilities and for port and harbor improvements, etc.

We recognize the fact that future federal funds might not be made available for dredging programs. Therefore, the coal industry must make further commitments to insure its destiny. In this regard, if the administration will fast-track its permitting and environmental policies, then we are prepared to pay our fair share of dredging expenses. Such a position has been adopted by the coal companies under the National Coal Association.

REVISION OF RAIL FREIGHT RATES

Another major item of importance pertains to rail freight rates. More cooperation and understanding is needed from our railroads in the setting of rates to tidewater and river ports. These rates, when coupled with the investments in terminals, vessel loading delays and demurrage make the total in-vessel cost for U.S. coals even less competitive with other foreign coals. We do not imply that our railroads should not be profitable or make a fair return in the pricing of their services. However, they need to provide some form of freight pricing stabilization to encourage exports and intermodal movements. Contract rates and tonnage guarantees would be one means to ensure carrier revenues.

IMPEDIMENTS TO ILLINOIS BASIN COAL DEMAND

Illinois Basin coals have historically and predominantly moved to, and served, the domestic, industrial, and utility markets--and to some extent, the metallurgical market--in the nearby midwestern area. This has been a good and dependable fuel source for this area and, to a great extent until the last 10 years or so, satisfied the area's productive capacity. Unfortunately, while U.S. production has increased in recent years, Illinois Basin coals have not enjoyed this increase in demand. I feel this has been due to several factors: 1) Illinois Basin coals are a bit lower in heat content than eastern coals, but, when cleaned, generally can be in the 12,000 Btu range; 2) Illinois Basin coals have generally moved by river and rail to markets within a several hundred mile range; 3) Illinois Basin coals have heretofore not moved to the export market due to their location in mid-America, and steam coals until very recently were not exported to any great extent except to Canada; 4) Illinois Basin coals have been greatly affected by the influx of gas in the metropolitan areas and the environmental re-strictions instituted by the EPA due to the higher sulfur content of most of the coals originating from this area. While there may be other reasons, I feel that these fairly well cover the reasons behind the static or declining use of Illinois Basin coals.

ILLINOIS BASIN RESERVES AND POTENTIAL

From a reserves standpoint, this area is a vital energy source, one estimated to have over 200 billion tons of identifiable and recoverable reserves. The area has a number of our nation's best and most-qualified producers; this area has a mature and stable labor force. With the growing demand for steam and industrial coal abroad and with a transportation network comprised of the Mississippi River and progressive railroads, it would seem that there is no reason why these coals of better-than-average quality

should not find their place in the energy-short areas of Europe and
the Far East.

HIGH SULFUR PROBLEM

The one and only reason why Illinois Basin coals would not be
marketable in the world market is the higher sulfur content of these
coals. Export coal buyers generally fall into four general
categories where sulfur specifications are concerned:

> under 1%
> up to 1.5%
> 1.5 to 2.5%
> 2.5 to 3.5%

Illinois Basin coal by itself cannot fit easily into the first two
categories. Some coals from this area might fit into the third
category, with more falling into the last category. The larger
percentage of utility and industrial coal demand now falls in the
third category--1.5 to 2.5% sulfur.

COAL BLENDING

As I see it, the best way for these Illinois Basin coals to fit
into the growing export market is by utilizing a blending procedure
at port in or around the New Orleans area. Lower sulfur coals from
the eastern area--up the Ohio--can be loaded by barge and assembled
with barges loaded on the lower Ohio or Mississippi River to meet
the required sulfur specification when loaded into vessels at the
port. Barges can be transloaded intermittently or loaded from
storage to ensure the needed quality. Many Illinois Basin coals can
be loaded directly to the river, thus reducing the cost of
transportation. One also has an excellent railroad system that
serves the area and the ports of the South. In utilizing the
railroad system, blending of coals would probably be most effective
by combining river movement of lower sulfur coals with higher sulfur
Illinois Basin coals at one of the terminals in the port area.
    The demand for steam coal overseas will be great in the years
ahead, and with the capacity that presently exists in this area for
producing coal, and with the transportation system that is already
available, and with proper coordination and planning, I can see no
reason why Illinois Basin coals should not find a more prominent
place in the export market.

We hear today much about our having an energy crisis. We in this country have no energy crisis. Our country is blessed with an abundance of energy or potential energy. We have an oil crisis; we are dependent upon foreign oil at whatever price it may be. We should make use of our coal resources. In conclusion, I feel that the coal industry can be optimistic about the opportunities for export coals and in attaining successful long-term marketing programs. We have much to gain if we react expeditiously to this rapidly developing worldwide coal demand. Finally, our nation, our state, and our industry must demonstrate to those here at home and throughout the world that we have the resources and capabilities to produce and supply both the needs of the U.S. and those abroad with our most abundant fuel source--coal.

ILLINOIS BASIN COAL
RAMESH MALHOTRA

Within the Illinois Basin, which includes coal deposits of
Illinois, Indiana, and the western part of Kentucky, over 215
billion tons of coal resources have been identified. Of this
amount, most recent estimates show that at least 90 billion tons
could be economically mined. Considering that currently only 140
million tons of coal per year are being produced from this
coalfield, it is apparent that current production could easily be
expanded by tenfold without experiencing any reserve availability
constraints.

In addition to the availability of large reserves, better
mining conditions enable coal in the Illinois Basin to be produced
at a relatively lower cost than Appalachian coals. Historically, on
the average, Illinois Basin coals, on an FOB mine price basis, have
been sold at anywhere from 20 to 25% below Appalachian coals. We
expect, as the more easily minable coal reserves in the Appalachian
field are depleted, the differential in the costs of production will
become even greater.

Because of its location, the Illinois Basin coalfield is
accessible via four navigable rivers and is also served by a network
of 10 different railroads. The accessibility of this coalfield by
such a diversified transportation network makes these coals a
dependable source of supply for both domestic and international
trades.

Another attractive feature of Illinois Basin coals is the
uniformity of the product that is shipped to the customer. In the
Illinois Basin, over 85% of the coal found is known to contain
sulfur exceeding 3%. To reduce the sulfur content, most of the coal
produced in the Illinois Basin is mechanically cleaned before it is
shipped to customers. Because of the mechanical cleaning, Illinois
Basin coal producers are able to provide their customers with a more
uniform product.

DISADVANTAGES OF ILLINOIS BASIN COALS

In spite of all these positive aspects, Illinois Basin coals
have not always been favorably accepted within the U.S. or in the
international markets. Domestically, federal, state and local
regulations are constraining the expanded use of Illinois Basin

coals. In the international markets, there are basically two reasons why the Illinois Basin coals have not yet been accepted. The first is a lack of knowledge about Illinois Basin coals within the international community, and the second factor is the stringent coal quality requirements that are being imposed by international coal buyers who are seeking U.S. coals for security and dependability.

Within the past year, due to the problems of ocean vessel demurrage and reliability of coal supply from Appalachia, Illinois Basin coals have come to the attention of several international coal buyers. Some coal from the Illinois Basin has already been purchased and shipped to Europe and Japan. It is expected as long as the problem of demurrage in the East prevails, Illinois Basin low sulfur coals will continue to find some markets in foreign countries.

Unfortunately, the quantity of low sulfur coal that could be supplied from the Illinois Basin is limited. We envision, due to this resource constraint, the expanded use of Illinois Basin coal in the international market will not take place until the foreign buyers modify their coal quality specifications to accommodate Illinois Basin high sulfur coals.

PEABODY COAL COMPANY REVIEW
WAYNE T. EWING

As President of the Illinois Division, I have the operational responsibility for the eight Illinois-based mining facilities, which last year produced over 11 million tons of Illinois coal. Some 60% of that coal was produced by underground operations, and the remaining 40% represents the division's three surface mines' production.

Currently, Peabody's so-called "Belleville Mining Area" controls approximately three billion tons of coal reserves in the state of Illinois. Over one-half of these reserves are located in the southwestern portion of the state in and around the Belleville Mining Area, which covers portions of St. Clair, Randolph, Washington, and Madison counties. Intersecting the Peabody reserves in this area is the Kaskaskia River. The Kaskaskia River is a channelized waterway which empties into the Mississippi River below Evansville, Illinois, in Randolph County. Also, crisscrossing this area is the Illinois Central Gulf, Missouri Pacific, Family Lines and Southern Railroads. The Belleville area presently serves as the center of Peabody's mining activities in the state and accounts for five of the Division's eight Illinois-based mines. It is also primarily in this area that the company is planning its future coal development efforts, which are partially under consideration in response to an expected expanding coal export market.

PEABODY'S PRODUCTION CAPACITY

In attempting to assess the division's productive capabilities in response to the increasing export market, I can offer the following scenario reflecting Peabody Coal Company's status as the nation's largest coal producer (in 1980, for example, Peabody produced over 59 million tons of coal from a 10-state area):

1. The division has at present an excess productive capacity of some 10% at its operations in the Belleville area. This means that, should conditions warrant, the five area mines could produce approximately 600,000 additional tons of coal during the second half of 1981. We also have the flexibility of adding additional production capacity at one of our mines in a relatively short time-frame.

2. The division has, in an advanced stage of planning and partial development, two additional underground mines for the Belleville area. These mines, known as St. Libory and Tilden, could be operational in a two- to four-year time-frame, and could provide, at full production, approximately 2.8 million tons annually. While these two mines were scheduled to be brought on line gradually to replace declining production at the area surface mines, they could, in fact, be brought on line much quicker in response to a strong foreign market and utilized until needed for tonnage for current contracts serviced by our depleting surface mines.

3. The division has in the planning stage a large mining complex to be located along the northern, channelized portion of the Kaskaskia waterway between New Athens and Fayetteville, Illinois. This complex will consist of four new underground mines, a new coal preparation plant, and a new dock facility on the Kaskaskia. Should market conditions warrant, initial coal deliveries from this potential six-million-tons-per-year complex could commence by the mid 1980s, with full production coming in the latter part of the decade. In summary, assuming favorable market conditions, the division's productive capabilities in the Belleville area, by the late 1980s, could be nearly double its present annual capacity.

PEABODY'S RESERVES

I would also like to point out that the division has other coal reserves in the state that could be allocated to the foreign market. Our Rileyville Reserves, in the far eastern portion of the state, contain strong coals, low in sulfur and high in heat value, which are presently under evaluation for potential export.

PEABODY'S TRANSPORTATION POTENTIAL

Not only does the Illinois Division have excellent productive capabilities which could provide significant coal tonnages for the foreign market, it also has a superior transportation capability in the Kaskaskia River. The location of the Division's reserves and mining facilities in relation to this waterway provides the company with a direct link to the Mississippi River, thereby giving us an excellent foreign commerce potential.

In concluding my remarks, I would like to relay Peabody's assessment of the export coal market, and news of the company's efforts to participate in this market.

RESPONSE TO A RISING COAL MARKET

Some time ago, Peabody Coal Company recognized the importance of the worldwide coal market and foresaw the United States as a leading supplier in that market. In view of Peabody's leadership in the United States coal industry, Robert H. Quenon, the company's president and chief executive officer, initiated an intensive program, which was designed to market Peabody coal abroad. Since that time a number of activities have taken place:

1. Coal reserves have been identified in a number of Peabody's operating divisions, and those reserves have been, and are currently being, evaluated for potential export.

2. Peabody sales and marketing personnel have been dispatched abroad to meet with potential customers. Selected foreign customers have, in turn, been invited to meet with Peabody personnel and to tour Peabody facilities. As an example, I might mention, that the Illinois Division recently hosted an important trade delegation from Taiwan and representatives from France.

3. We have been working with other companies concerning the construction of a large coal loading terminal in the New Orleans-Baton Rouge, Louisiana, area which could play an integral role in Peabody Coal Company's efforts to market its coal abroad. A decision to proceed with construction of such a facility is expected before the end of this summer.

4. Peabody has also publicly supported the Corps of Engineers' proposed project that would provide a deep draft (55-foot) channel to the ports of New Orleans and Baton Rouge, Louisiana, and to other port facilities on the East and West Coast. Peabody Coal Company is currently engaged in numerous other activities designed to promote coal exports. The areas of involvement I have mentioned here today should give you an indication of our overall commitment to this vital and growing market.

OLD BEN COAL COMPANY REVIEW
J. HARLEY WILLIAMS

As in the nature of all economic projections, the forecasts set a range for U.S. coal exports in an attempt to describe the potential of an uncertain event. Witness the projections developed by the World Coal Study team and the Interagency Coal Export Task Force. While the absolute numbers themselves are of value, the range of the projections and the description of both potential and uncertainty are more important. As I will elaborate later, the issue of range is key to a discussion of the potential for Illinois high sulfur coal exports.

The events of a balanced fuel program the past several years have finally given us an understanding of the fact that we are dealing with world resources in a world economy, and that no mineral energy resource is infinite regardless of how abundant it appears at any given point in time. Had we in the United States had the foresight to come to grips with this fact years ago, we never would have advanced the view that coal was the "fuel of last resort" in meeting many of our electrical and industrial energy requirements. In a finite system, no resource should be viewed as a last resort if a balanced energy program is ever to be achieved. Because coal was viewed as such during the era of perceived oil and gas abundance, we found ourselves in the 1970s burning several million barrels per day of oil equivalent as boiler fuels, a role that should have been performed through extensive coal utilization. But, the objectors were strong and the objections to increased coal use were many and varied. As a result, while biased incentives to use oil and gas (with low controlled prices) were kept in place, oppressive regulations disrupting coal use were imposed, and the high sulfur coal segment of the industry was particularly hard hit. So, it is ironic, yet interesting, to note that we are addressing the means to use Illinois high sulfur coals overseas, while the state's largest electric utility burns only token amounts of Illinois coal.

SHORT-TERM VS. LONG-TERM RESOURCE MANAGEMENT

There is, however, a message in our history of taking the short-term view of resource management, and a lesson to be applied to the development of Illinois coal exports in the future. While we recognize the need for staged advancement of resource management, it should not be necessary to develop a stage that retards the long-term goal of a balanced energy program; instead, a program that advances the utilization of all coal resources, regardless of sulfur content, is in order. It is our job to ensure that foreign countries do not develop the short-term view that while coal burning has become necessary, high sulfur coal burning is now the "choice of last resort." This viewpoint is regressive and will not provide sustained growth across all coal markets and all coalfields. Illinois coal operators cannot accept a "boom and bust" environment. To commit the necessary capital and manpower to an integrated export development program, sustained and assured growth is vital. Sustained growth minimizes the uncertainty and will bring us closer to achieving the upper range of recent coal export projections. The fate of Illinois high sulfur coal exports lies here.

European and Pacific Rim countries have become as environmentally sensitive as the United States. The advancing view is that the current availability of world low sulfur coals will be adequate to sustain primary energy needs under most low-growth scenarios. The fact that non-U.S. low sulfur coals are more than competitively priced even as compared to our high sulfur coals, creates a strong impression of ample resource availability. High sulfur coals are seen only as a necessary evil when higher demands ultimately bring us closer to the upper range of coal export projections. The underlying assumption is that if the low sulfur coals are abnormally depleted, the higher sulfur coals will be there as a fallback measure. The prudent resource management decision would be to stretch out those low sulfur resources for selected industrial processes in environmentally sensitive areas.

DIVERSIFIED COAL USE

To achieve a sustained growth for Illinois coal exports, we should help overseas users understand that the utilization of a wide range of coals in environmentally acceptable ways will serve both their short-and long-term interests best. We must find niches in the export market where our coals fit best. Then, we must be cost-effective in our mines and transportation systems. We must work on engineered sales, sell our ideas overseas, and, we must "think big." A balanced commitment to coal use is needed, including power generation, industial processes, and, in the near future, supplementing of liquids and gases through synthetic conversions. Ultimately, coal should be most valuable for its carbon.

A BALANCE OF FUELS

Each energy resource--oil, gas, uranium, oil shale, tar sands, solar power in all its forms, low sulfur coal and high sulfur coal--has its own optimal, end-use application: coal, uranium and geothermal for boiler fuel; coal for carbon; oil (natural and synthetic) for transportation; gas (natural and synthetic) for chemical feedstocks and residential use; and solar, wind, hydro, and biomass, for supplemental heat and energy. The objective is to ensure, both here and abroad, that the proper long-term application is realized, and wasteful, short-term solutions are discouraged. This task should begin now, at home.

ILLINOIS EXPORTS

The Illinois coal industry is poised to provide reliable export supplies. The state ranks first in recoverable underground reserves and third in total strippable reserves, behind only Montana and Wyoming. There is a good mix of underground and strip productive capacity currently in place. Because many operations have been running with reduced shifts or shortened workweeks, there is substantial surge capacity readily available on short lead times. Illinois reserves are held by strong, diversified corporations with the capability of providing capital resources for the long-term, and the newest entrants in the industry will add even more strength to the development of the state's resources. In addition, Illinois mines, both deep and strip, are highly mechanized and safe, with modern preparation plants and coal loading facilities.

Illinois mines appear to have weathered the damaging impacts of the Federal Coal Mine Health and Safety Act of 1969, as amended in 1977, and many other regulations, and a turnaround in productivity has begun to take place. Between 1969 and 1979, Illinois underground mine productivity was more than cut in half--from 22.9 tons per-man-day in 1969 to 11.2 in 1979. For the year 1980, productivity increased to 13.5 tons per-man-day, a 20% recovery over the earlier year. We expect the trend, not necessarily at the same rate, to continue into the 1980s. Illinois surface mines have employed sound reclamation programs over the years, and should not have to play substantial catch up to comply with new reclamation programs.

INFLUENCE OF THE UMWA

The highly concentrated United Mine Workers of America, a labor force in the state, has caused considerable work stoppages, but we have hopes that this will change. Under the 1978 UMWA contract agreement there have been 85 to 90% fewer work stoppages in the state compared with the previous agreement. The ratification of the initial 1981 contract proposal by Illinois' union membership, points toward a revitalized work ethic and an understanding of the need for reliable labor in the coal mines. It is most unfortunate, though, that the present contract strike is with us today. We believe this union is doing unbelievable damage to its members and itself and to those operators employing UMWA labor.

ILLINOIS TRANSPORTATION

The transportation options available to Illinois coal producers are second to none in the country. An excellent railroad network is in place providing good, all-rail access south to the Port of New Orleans and north to the Great Lakes an St. Lawrence Seaway System. Short rail hauls from the coalfields to river terminals on both the Ohio and Mississippi Rivers for barge movement beyond should provide healthy competition between transportation modes.

We sincerely believe that private industry is the proper way to handle the development of U.S. coal exports, and we would ask government only to support us in our objectives. The state's coal resources, productive capacity, capital requirements and infrastructure are sound. What is needed is our own initiative to cultivate an emerging market overseas, while reaffirming our commitment to a balanced resource management at home. We are willing, too, to keep the lines of communication open between industry and government, so that if problems are encountered, a concerted effort can be made to remove procedural hurdles.

ZEIGLER COAL COMPANY REVIEW
MICHAEL REILLY

Illinois is very fortunate in many ways. We have an estimated 162 billion tons of bituminous coal, of which 30 billion tons are considered to be economically recoverable at this time. We have many of the finest coal miners in this country and many other men and women who would love to have a job in mining coal. We have the finest transportation network in the nation with excellent highways, numerous railroads, and easy access to the inland waterways. We have established companies with many years of mining experience and the ability to raise the necessary capital to build new mines.

The major problem is that the sulfur content of most Illinois coal is higher than what is presently being specified for many foreign markets. Most export markets are looking for coal with a sulfur content of less than 2%. These low sulfur specifications have in many cases been arbitrarily set and are not actually required. Also, low sulfur coal has been available from other states.

The increased demand from the export market has seriously strained the coal-docking capacity at our eastern ports. Many foreign ships are waiting to be loaded with coal at these ports and this has caused a dramatic increase in the cost of coal to the customer. This gives Illinois an excellent opportunity to help supply this growing market. Coal transfer facilities are being expanded in the New Orleans area. This is the final link in the network needed to move Illinois coal from the Mississippi to the Gulf and on to Europe and Asia.

Another factor in the international utilization of high sulfur coal is cost. Illinois coal lies in thicker seams and in most cases can be mined at lower costs than eastern coals. We have a substantial transportation advantage over western coals. We should be able to deliver coal at lower costs to the export markets and this may cause a reconsideration of sulfur specifications.

Another factor in the international utilization of high sulfur coal is the dependability of supply. There are presently several million tons of excess capacity available from Illinois mines. Most Illinois mining companies have long records as dependable suppliers, and with the availability of excellent human resources will continue to be dependable suppliers.

Creative marketing ideas can also help to increase Illinois' share of the export market. To help overcome the sulfur problem, Illinois coal can be loaded at the transfer terminal with low sulfur coals. We can offer to help with the technology and equipment to reduce the sulfur dioxide. We can make it cost effective to burn Illinois coal on applications that do not require low sulfur coal. Working together, we can find ways to increase our share of this international market. Illinois has the coal reserves and the men and women to mine it and is equipped to deliver it to the Gulf for shipment to foreign markets.

CONSOLIDATION COAL COMPANY REVIEW
J. ALAN COPE

Consolidation Coal Company is the nation's and Illinois' second largest coal producer. In 1980, production from Consol's mines and those of others supervised by Consol totaled 49.6 million tons. Consol's production in Illinois was 8.6 million tons in 1980. Consol holds 1,774 million tons of reserves in the Illinois Basin, of which 1,066 million tons are in the state of Illinois. On a clean, dry basis, washing at a specific gravity of 1.6, the average sulfur value for Illinois Basin coal is 3.0%, or 4.5 pounds per million Btu. On the same clean, dry basis, the average sulfur values for Illinois coal is 2.8%, yielding 4.3 pounds per million Btu.

Consol thus has a strong interest in the mining and use of Illinois Basin high sulfur coal. The most logical place to market this coal would be in Illinois itself and adjoining states. However, the Clean Air Act has severely impaired the marketability of the coal in this region. We estimate that 10 to 12 million tons of Illinois Basin coal have been displaced by western coal because of the SO_2 standards imposed pursuant to the Act. In our view, the standards were set without regard to the scientific basis for the limits imposed or their cost effectiveness.

THE CLEAN AIR ACT

There will shortly be an opportunity to amend the provisions of the Clean Air Act, and the nature of these amendments could have a marked effect either for better or for worse on the marketability of Illinois Basin coal. It is hoped that no steps will be taken--especially in view of the adverse effect on employment and the economy--which would lead to further coal imports into the region which is richly endowed with coal resources and should be a net exporter rather than a net importer of coal and energy.

Partly because of the precedent of U.S. sulfur restrictions, it is not surprising that, as we have restricted the use of Illinois coal in its home region, we find that other countries are reluctant to accept it. Consol has recently prepared a survey of the European steam coal market. It is estimated that European steam coal imports will increase from a 1980 level of approximately 83 million metric

tons to 96 million tons by 1985 and 165 million tons by 1990. In
that year, we estimate that only 7% of the import requirements will
be for coals of a higher sulfur content than 2%. This would seem to
limit the market for coal of the type found in the Illinois Basin to
about 12 million tons in Europe in 1990. However, this market would
have to be shared with Northern Appalachian coals which have a
transportation advantage to Europe.

These economics exclude consideration of demurrage caused by
East Coast port congestion which is currently a substantial adverse
factor in the economics of shipment of coal from the East Coast of
the U.S. and can add more than $10 a ton to the price. However, by
1990, we expect sufficient port capacity to be installed so that
congestion and demurrage will no longer be a factor, thus reducing
the relative attraction of New Orleans and the Great Lakes ports as
alternatives to the East Coast.

PROSPECTS FOR ILLINOIS BASIN COAL

Although this is an admittedly bleak picture for high sulfur
coal exports, it is possible that it could change. Western Europe
has been able to obtain low sulfur coal from its own production and
from such exporting countries as Poland, South Africa, and
Australia. It has also been able to obtain low sulfur coal from the
United States. However, as we approach the 1990s with an increase
in the volumes consumed, significant premiums may develop for low
sulfur coal in order to shift some volume from the metallurgical
coal market to the steam coal market. This could be enhanced if the
Clean Air Act is amended to reduce the scrubbing requirements for
low sulfur coal for new plants. This low sulfur premium may
encourage European buyers to look at the blending of coals to reduce
cost. For example, it may be feasible to blend Illinois Basin coal
in LeHavre, Rotterdam, or other large import terminals.

A second development which could increase the acceptable level
of sulfur content would be the installation of scrubbers on utility
plants in Europe. In the Federal Republic of Germany, a new utility
plant of more than 400 Mw will be required to install scrubbers.
However, these may be one-stage scrubbers designed to accept only
1.5% sulfur coals, and this would not materially assist the
development of markets for higher sulfur coals. We hope that
European buyers could be encouraged to increase their supply
flexibility by installing desulfurization equipment, which would
permit them to use a wider variety of coals.

Another development which could lead to the acceptability of
high sulfur coals would be the introduction of fluidized bed

combustion technology. A Consolidation Coal Company affiliate, the Conoco Coal Development Company in conjunction with Stone and Webster, has developed a process in the pilot plant stage which can remove 90% of the sulfur contained in the coal. This plant is most suitable for industrial application and we hope to install a commercialized version in the near future. This process and similar ones which other companies are working on could expand the market for high sulfur coal in Europe and in the United States. However, the effect of the process on volume requirement will not be felt until the 1990s and then principally in the 25% of the market which is non-utility use.

Another possibility for expanding high sulfur coal exports will be the introduction of gasification processes which could be used for the manufacture of natural gas substitutes and methanol. There is likely to be a market in the 1990s with perhaps some initial units being built and brought into operation before 1990. While such units may utilize 1.5 million to 3 million tons a year each, it is unlikely that more than a few will be built in Europe during the next 15 years. A variant of this possibility would be to establish methanol plants in the United States to export the methanol instead of the coal.

COAL IN EUROPE IN THE 1990s

The role of coal in the European energy picture in the 1990s will depend on many factors including the relative price relationships which develop for coal and oil, the future of nuclear, and the development of technology. A key factor in the equation will be the ability of the United States to provide coal at a competitive price and make it available through efficient ports dredged to accept large vessels and built with sufficient capacity to avoid demurrage. It will be necessary for the United States to keep its costs under control with respect to severance taxes, environmental regulations, waterway and rail transportation rates, and labor and other production costs. Given a favorable environment, the European import requirement could grow to between 250 and 300 million tons per annum by the year 2000. The import of coal in this volume will require the acceptance of higher sulfur coals. By that time, desulfurization technology will have advanced. Moreover, at this point the volume of U.S. exports will have increased to a level at which a substantial percentage will need to come from the Illinois Basin through the Gulf ports. Thus, the prospects of high sulfur coal exports from the Illinois Basin in the late 1990s would appear to be encouraging.

ALTERNATE MARKETS

I have concentrated on Europe in this review because this market will be by far the largest steam coal export market for U.S. coal. There will, however, be some small volumes required in the South American market and the Pacific Rim countries in the Far East. These are estimated to require 100 million metric tons in 1990. However, it is unlikely that more than a small volume of this will be U.S. coal because of the competition from closer sources, such as Australia and China, and from Canada, South Africa, and Colombia. Some 5 million tons of U.S. coal may move from the West through Long Beach, and some marginal volumes will move out of New Orleans, though not more than 2-3 million tons. Because of the low sulfur content of much coal reaching Pacific Rim countries, it is possible that some of this could be high Btu, high sulfur coal for blending.

In order to advance the date at which substantial export volume from the Illinois Basin occurs, the United States government can take steps to influence the outcome. First, it can set an example to the rest of the world by basing air pollution standards on the best available scientific evidence. The Clean Air Act needs to be modified and streamlined with this in mind. Second, government can support the development of desulfurization technology, including fluidized bed combustion technology. Third, ports should be dredged to permit the use of large vessels to reduce the cost of transporting coal to foreign markets.

V

Transportation of High Sulfur Coal
to Export Ports

RAILROAD TRANSPORT
HARRY T. BRUCE

Much has been said of export coal in recent months, but little of this attention has focused on the Illinois Basin and its coals. The Senate Subcommittee on Energy is to be commended for its recognition of the potential contribution of the Illinois Basin both to reducing the U.S. balance of payments deficit and alleviating the world's energy situation, as well as its benefit to the economic prosperity of the area. At the Illinois Central Gulf (ICG), we believe strongly in the export potential of the Illinois Basin's enormous coal reserves.

As anyone who has read the MIT World Coal Study knows, world trade in steam coal is expected to explode from 73 million tons in 1978 to as much as 800 million tons by the turn of the century. The Illinois Basin has 215 billion recoverable tons of coal reserves, largely steam coal. It represents one of the largest reserve areas in the United States, and the majority of these reserves are served by the ICG.

The ICG sees coal--especially export coal--as one of its biggest growth commodities. To many, that statement may seem ironic coming from what is sometimes erroneously thought of as a north/south "granger railroad," whose role has been limited to carrying coal from the mines to midwestern utilities or to the nearest rail/barge terminal. Unlike other major coal carriers--and we are one of the 10 leaders--ICG's main line parallels the Mississippi River, which carries the nation's heaviest barge traffic. Conventional belief in the cost efficiencies of barge transport would seem to dictate that ICG leave export coal to the river and concentrate on domestic coal and other commodity movements.

But traditional thinking--at least, traditional rail/coal thinking--is going to change. The Staggers Rail Act is beginning to change many of the historic railroad commercial habits. The rails now have the full range of freedoms and flexibilities with which to innovate and to compete with barges, or any other mode of transportation for that matter. Successful marketing of U.S. coal abroad demands creative, aggressive methods. It isn't enough to jump on a plane, fly to Europe, and tell a coal buyer that you can

give him great turnaround time on unit train coal movements. Nor is
it enough for coal companies, with limited preparation, to whisk
piles of contracts to Europe with little thought as to the needs and
characteristics of the international market. The name of the export
market for railroads and coal companies alike is "landed Btu's."

As with any change, this one also requires a "learning curve."
This learning curve will include altering some well-entrenched
attitudes. This is particularly true when we talk about exporting
Illinois Basin coal since its viability in international markets is
surrounded by a peculiar mythology, considered fact by many.

ICG PLANS FOR DELIVERY OF COAL

Before debunking these myths, let me provide a frame of
reference for my remarks and the ICG's optimism regarding the export
market. Any discussion of any coal in the international marketplace
is really a discussion of "landed Btu's." Thus, any evaluation of
the international opportunities for Illinois Basin coal must include
a thorough evaluation and assessment of each of the steps involved
in delivering those Btu's--a systems analysis of exporting Illinois
Basin coal.

I should point out here that, unlike many transportation companies, the ICG owns no coal mines. The railroad's interest in the
export market is simply to determine how the ICG might participate
in the mine-mouth to ship-berth transportation of coal destined for
export.

Each of the steps leading to the ultimate coal consumer has
been subjected to systematic analysis by the ICG. Beginning with
the characteristics and requirements of the export market, one must
then blend in other steps important to a systematic approach:
inland transportation options, port facilities for transshipping the
coal, ocean transportation, destination ports, and the
transportation options within the purchasing country.

Overkill? Not at all. The ICG first had to determine whether
there was a profitable link in the distribution chain for the
railroad. The customer, who is buying delivered Btu's, is financing
the capital structure of the entire distribution chain. The ICG
could provide the best, most economical service from mine-mouth to
ship-berth, but if the shipment must also go to a destination
overseas beyond our control, that last link is crucial to the
viability of the whole chain. Or, still, if the coal doesn't meet
the buyer's specifications, the best transportation system in the
world isn't going to convince the importer to buy.

Obviously, the ICG has determined that exporting Illinois Basin coal represents a major opportunity for the ICG. The ICG has devoted a great deal of time, money, and human resources to making sure that opportunity is real, rather than an export fantasy engendered simply by reading the World Coal Study. The railroad believes that its studies of the situation are sufficiently thorough and the results sufficiently realistic to soundly debunk the mythology mentioned earlier. Let me take you through each of the myths and show you the facts the ICG analysis uncovered.

ILLINOIS COAL AND EUROPEAN NEEDS

First, there is a common belief that Illinois Basin coal will not satisfy European coal buyers' specifications. The fact is that Illinois Basin coal is acceptable. To determine the accuracy of the belief, ICG began collecting available data that would either corroborate or refute the statement. The market analysis began with MIT's World Coal Study, the first authoritative assessment of the world market and its needs. But the WOCOL study is a macroscopic view of the world marketplace and merely provided clues to specific areas that were examined in greater detail. To focus more clearly on specific international opportunities for Illinois Basin coals, the ICG conducted its own study of the Basin, its coals, and probable international sales prospects for those coals. (Excerpts of this study appear in Appendix A.)

THE ICG AND HIGH SULFUR COAL

Some of the Illinois Basin coal served by ICG is higher in sulfur content than is generally acceptable under current U.S. federal regulations. But that isn't necessarily a major deterrent in dealing with exports to Europe. The term "high sulfur" begs for closer examination. High, compared to what? One could go on at great length discussing the merits of one nation's or one region's steam coal properties. That is not what is important. What is important are the international market specifications. And sulfur content is but one of several key specifications that go into the demand equation. The important point is that there are markets abroad seeking coals which coal suppliers in the Illinois Basin can readily satisfy.

ILLINOIS BASIN COAL AND SPAIN

Currently, ICG is fine-tuning its focus on Europe even further by developing in-depth studies of prospective buyers on a country-by-country basis. The first to be completed is Spain. The results of that study indicate that Illinois Basin coals are both usable and acceptable, sulfur notwithstanding, to both the private enterprise utility and cement industries in Spain. Cement industry sources indicated that coal with up to 2.5% sulfur is readily acceptable. In the utility industry, there are several environmental initiatives under consideration that will make the Illinois Basin coals viable in this market.

The Spanish study didn't stop with determining the acceptability of Illinois Basin coals. It also pinpointed specific utilities and cement plants--both existing and planned--port facilities, coal demand, and import projections, Spain's infrastructure for receiving and handling coal, coal procurement procedures, and an evaluation of the National Energy Plan of Spain. The ICG believes that it has verified the existence of an Illinois Basin coal market in Spain. The ICG has conducted a similar study of the utility and cement industries in Italy, and has commissioned a study of potential Greek markets. The results of these studies are being made available to interested Illinois Basin coal producers.

RAIL VS. BARGE TRANSPORT

Since there is an international market for Illinois Basin coals, let's look at the next link in the export chain--transportation from the mine-mouth to the ship-berth. This link also is surrounded by myth, namely that, "Illinois Basin coal must move by barge to Gulf ports." Perhaps this myth has persisted because river water looks cheaper than welded steel rail. The fact is that the ICG can be totally competitive with barge transport from mine-mouth to ship-berth.

The natural tendency is to look at barge rates and rail rates, see the lower figure on the barge side, and say, "We better go by barge; it's cheaper." What those barge rates don't show on the surface are a number of "hidden costs." For example, the barge rates don't include the mine-to-river transport costs, nor do they include the extra handling costs or product loss caused by transfer from rail or truck to barge. They also do not include heat value losses associated with lengthy exposure to snow or rain. In contrast, fast, efficient, unit train movements are reliable regardless of low water levels, ice jams, or congestion.

THE STAGGERS ACT OF 1980

More important even than these factors is the flexibility the Staggers Act of 1980 has provided to aid railroads in developing and pricing the inland transportation segment of the coal export chain. This new flexibility has given us the wherewithal to work with coal producers and prospective importers to develop an energy delivery package that is beneficial to all concerned. The Act legitimized contract rates; it established new, more flexible pricing procedures that enable railroads to respond quickly to competition in the marketplace. For example, just prior to the passage of the Staggers Act, the ICG secured a new flexible tariff that permits it to change price on coal movements from Illinois to the Gulf on two days' notice to any point in a range between $16.50 and $11.25 per ton. This allows ICG to position itself relative to barge competition in the market. Because the ICG secured the tariff before Staggers, we are the only railroad that has it. Of course, barges are talking in excess of $6 to $7 per ton, but remember the hidden costs I mentioned a moment ago. The rail link in the chain represents one price from mine-mouth to ship-berth, all inclusive.

As you might imagine, the ICG took an especially hard look at this link and the economics involved, since it governs the profitability of ICG export coal movements. I can say without fear of contradiction that the railroad believes there is an exceptional opportunity for the ICG and its customers in mine-mouth to ship-berth coal movements. We need only debunk the myth by shipping our first unit train of Illinois Basin coal to the Gulf for export.

PORT FACILITIES AT THE GULF

The third link in the export chain centers on port facilities, and it, too, is surrounded by myth. This myth says, "There is no Gulf port facility to transload coal from rail to ship." This one has become myth only recently. Until five months ago, the statement was essentially true. In mid-January, however, an operating contract for the Public Bulk Terminal in New Orleans was awarded to Ryan-Walsh Stevedoring Company. I'm proud to say that our testimony on behalf of Ryan-Walsh and the export potential of Illinois Basin coal helped revive the Bulk Terminal. This facility is fully operational. The renovation work on the terminal permits it to handle readily up to 4 million tons of Illinois Basin coal per year. The Ryan-Walsh Bulk Terminal has an unloading rate of approximately 1200 tons per hour, blending capabilities, ample ground storage, and rapid ship loading--also about 1200 tons per hour.

But the Ryan-Walsh Bulk Terminal will provide only interim rail to ship transloading. Its limited capacity is insufficient for the amount of coal that should be moving into international waters from the uncongested, warm weather Port of New Orleans. Nevertheless, it does give Illinois Basin coal an efficient, economical entry to world markets. Currently, several organizations are looking at New Orleans area sites for a modern, rapidly unloading, coal-handling terminal with an annual export coal capacity of at least 20 million tons. The ICG is doing everything it can to help make that facility a reality as soon as possible.

COLLIER SERVICE

The fourth link in the chain is collier service to the importing country. Again, it is shrouded in myth, one which says, "Ocean freight from New Orleans will be more expensive." Implicit in this statement is the assumption that more expensive means prohibitively more expensive. As pointed out earlier, the discussion is really one of landed Btu's.

EAST VS. GULF COAST OCEAN TRANSPORT

First, ocean freight from New Orleans is only slightly higher than that from East Coast ports, exclusive of demurrage. However, extraordinary congestion at eastern ports makes additional Gulf coal handling capacity all the more appealing to the international coal market. Certainly, the demurrage charges--conservatively, $7 to 10 per ton--incurred as ships wait offshore for weeks on the East Coast more than offsets the slightly higher ocean freight rates from the Gulf to European destinations. Not only does the use of New Orleans as the exit port eliminate the monumental East Coast demurrage charges, but it saves weeks of delay in shipping time, thus improving ship utilization. Outlined in Table 1, are some of the estimates ICG developed when it assessed this link in the chain.

Table 1
LANDED BTU COST

	Via Gulf Coast	Via East Coast
Coal price	25-30	28-33
Rail rate	12	13
Rail to vessel	2.50	.40

	Via Gulf Coast	Via East Coast
Ocean rate	12-14	11-13
Demurrage	-	7-10
Unloading	2	2
Total landed cost per net ton	$53.50 - 60.50	$61.40 - 71.40
Landed cost per million Btu	$ 2.33 - 2.62	$ 2.56 - 2.98

The ocean freight difference is estimated to be between $1-2 per ton, favoring the East Coast. However, when all factors are considered, the situation shifts dramatically, putting the Gulf Coast in a more favorable light. The estimated cost per million landed Btu's in Rotterdam from East Coast ports came to between $2.56 and $2.98. From the Gulf Coast, the estimated range is $2.33 to $2.62. As you can see, ocean freight is not the proverbial weak link in the export chain.

EUROPEAN PORT FACILITIES

The two final links in the chain are port facilities and the transportation systems of the importing countries. These are the only two of six links not shrouded in myth. Nevertheless, they both require careful examination in any assessment of market potential.

Our Spanish study examined both these subjects during the basic market analysis, and the results supported the existence of a Spanish market for Illinois Basin coal. For example, several of the new utilities being built in Spain will be very close to existing or planned port facilities. Current port facilities are adequate to handle Spain's import coal needs for the time being. New facilities are planned to provide the necessary additional capacity to handle future needs. The existing transportation infrastructure in Spain is also adequate to supply both cement plants and existing utilities with coal transportation services.

Preliminary information on other prospective European importing countries has yet to reveal weakness in either of these links in the chain. However, there is a problem that must be surmounted before the Illinois Basin exports begin.

EUROPE AND APPALACHIAN COAL

Traditionally, Europeans who have purchased U.S. coal have purchased Appalachian coal and it has been shipped through our eastern ports. The coals have been adequate and the costs competitive with other world sources, such as South Africa, Australia, and Poland.

In conducting its studies of the European markets, ICG found that only Appalachian coal is thought of when U.S. coal is mentioned. Europeans see the eastern United States as a coal supplier, but see little beyond eastern Kentucky. For the most part, Europeans have heard little about Illinois Basin coals. The information they do have has come sporadically in bits and pieces, usually from a visit by a coal executive selling his company's products (see Appendix B). It is as if the mountains of eastern Kentucky form a wall or barrier that obstructs the European view of the Illinois Basin. (Unfortunately, this myopia extends even to the Federal Railroad Administration, which, for the most part, ignored potential Illinois Basin coal export in its projections of "Railroad Freight Traffic Flows 1990.")

As may be obvious, the most pressing need that must be met before Illinois Basin coal exports become a reality is a massive effort to educate and inform. This effort must be coordinated among railroads, coal companies, and ports. For its part, ICG has acted, and will continue to act, as a catalyst and strategist in bringing together potential importers and coal producers by the following:

1. speaking out in behalf of the Illinois Basin as opportunities arise;

2. providing informational materials, such as the "Import Coal Guide" which is now being prepared;

3. circulating the results of our country-by-country market studies to interested Illinois Basin Coal producers;

4. working with ports and coal producers to develop transportation price packages attractive to potential importers; and

5. supporting efforts to modernize as well as develop port facilities to meet future needs.

DEEP DRAFT ACCESS TO THE PORTS
OF NEW ORLEANS AND BATON ROUGE, LOUISIANA
HERBERT R. HAAR, JR.

To cite but one important export example, the growth of U.S. grain exports over the past 10 years is a phenomenal success story--a marvelous example of the free enterprise system at work. During the last decade, world grain trade just about doubled--rising from 108 million tons to 213 million tons. During the same period, U.S. exports increased from 46 million tons to 127 million--a gain of more than 250%. The U.S. share of world trade has risen from 40% to almost 60%. This is an example of growth both in real terms and relative terms. During the same 10-year period, Canada's exports remained unchanged.

The 10 modern export grain elevators between Baton Rouge and the Gulf handled 39% of all the nation's grain export in 1980. Clearly, our economy has benefited in terms of foreign exchange earnings and the creation of jobs. In fiscal 1979, agricultural exports amounted to $32 billion, or 20%, of this country's exports. Agriculture had a favorable balance of trade of $16 billion whereas the country had a deficit balance of $28 billion in total--clearly demonstrating the importance of grain exports as the major dollar earner. Conservative projections by the grain exporters show that by 1990 this grain movement alone will represent 96 million tons annually.

HANDLING FACILITIES AND CAPABILITIES

Producing grain for the expanding world demand would not have resulted in the huge growth of exports had there not been commensurate growth in handling facilities and capability. Continental Grain alone is investing a quarter of a billion dollars in rebuilding its port of New Orleans terminal. Huge private investments have been made in export terminals, interior loading stations, hopper cars, and barges. It is only natural that much of this capital has been invested in facilities from Baton Rouge to the Gulf on the Mississippi River. The river is the cheapest link between the main producing areas of this country and the major foreign markets. What is missing is the ability to get efficient

bulk carriers, 150,000 deadweight vessels, to these modern export
grain elevators.

PROPOSED CHANNEL

The project to provide a 55-foot channel from Baton Rouge to
the Gulf has strong local support. The President of the United
States and former President Carter have both spoken in favor of the
project. The Louisiana congressional delegation is strongly in
favor, and Senator Johnston has prepared legislation to expedite the
required review and funding. It is uncommon to find a waterway
project with such a high benefit-to-cost ratio--8.5 to 1--based
solely on the 1976 cargo movements. This benefit-to-cost ratio
could be doubled if the port's projected coal exports were included.

PORT EXPORT PREDICTIONS

The port is in a position to see, on a day-to-day basis, the
trends in international trade and the world markets available for
U.S. goods. The port's trade offices are located strategically
throughout the United States and the world. From reports received
within the past year, we are able to say unequivocally and without
hesitation that the future is very bright provided the deepening can
be accomplished on an expedited schedule. The port's projections
for coal movements involve extremely rapid growth--from one and 1.5
million tons in 1979 to 7 million tons in 1980, and an estimated 12
million tons in 1981, to 60 million tons before 1985, and double
that figure by 1990. The basis for predicting a rapid growth in the
use and export of coal is based on fundamentals of the energy
crisis. The cost of petroleum products continues to increase. In
the past two years, the cost of petroleum has more than doubled, in
part because the revolution in Iran reduced the export of petroleum
from that country.

ALTERNATIVES TO OIL

This nation and allied nations are deeply concerned with the
availability of petroleum and are seeking alternate sources of
energy. Hydropower is limited and, in the developed nations, has
been extensively exploited. Solar, wind, and geothermal are not
expected to be of significance until after the year 2000, and even
then will represent only a very small percentage of energy required.

Nuclear power has broad and strong environmental opposition. Only coal has the immediate potential to alleviate this nation's and allied nations' dependence on petroleum.

COAL EXPORT PROJECTIONS

WOCOL projects export potential of 350 million tons per year of coal from the United States by the year 2000. Dr. Ulf Lantzke, Executive Director, International Energy Agency, at the 63rd annual meeting of the National Coal Association, projected 100 million tons of coal export by the United States in 1990, rising to 300 million tons by the year 2000. He also projected that western Europe would be importing 500 million tons of coal per year by the end of the century. France alone is today importing over 35 million tons of coal per year and is seeking to purchase 10 to 15 million tons per year from the United States by 1985.

The Coal Exporters Association is predicting exports of 90 million tons of coal for 1980. This is only 11% of the U.S. total coal production of 815 million tons, and most of this export coal is metallurgical coal. The big demand in the future will be for steam coal, very little of which is presently exported from the United States. Of the 90 million tons, Hampton Roads is expected to ship over 60 million tons in the future from its piers at Norfolk and at Newport News. An average of 50 ships is waiting to load coal at these ports, and in recent months the backup has been as high as 100 ships and over.

LOWER MISSISSIPPI TRANSPORTATION

The tremendous interest in the lower Mississippi River, between Baton Rouge and the Gulf, is directly related to the ability of the Mississippi River to transport large volumes of coal. The barge lines recognize this, and two barge lines--the American Commercial Barge Line and the Federal Barge Line--have invested over $55 million in modern, efficient rail-to-water transfer facilities capable of handling 30 million tons of coal per year. The barge lines can transport this coal coming from the states bordering on the upper Mississippi River and its tributaries to the planned coal export terminals in units of 600,000 tons each. Additionally, coal slurry pipelines with a potential annual throughput of approximately 100,000,000 tons are being planned for terminals in Houston, Baton Rouge, and Florida. The one for Baton Rouge, with a 25,000,000 ton annual capacity, should be the first to come on-line.

PORT TERMINALS AND FACILITIES

At the International Marine Terminal, 50 miles below New Orleans, ample ground storage is available; barges do not wait for ships and ships do not wait for barges. This terminal will be expanded to handle 12 million tons of coal by 1982. Across the river, the Electro-Coal Transfer Company is expanding its capacity with a $100 million-plus two-phase expansion program. This expansion will permit it to handle 25 million tons of export by 1983. Near Baton Rouge, the River and Gulf Transportation Company has acquired almost 600 acres of land for an export terminal capable of handling 11 million tons of coal and 5 million tons of iron ore by 1985 and 15 million tons of coal by 1990. Five other coal terminals are in the planning stages and have either found sites or have been authorized to obtain options on land in addition, one recently rehabilitated terminal is now going on-line.

EUROPEAN IMPORTERS

In France, the government has a monopoly on coal production and purchasing. The Association Technique De L'Importation Charbonniere (ATIC) is the sole agency responsible for importing the large quantities of foreign coal necessary to provide replacement for oil. ATIC is also acting as agent for the Spanish and Netherland governments and is coordinating its efforts with West Germany. The aim of ATIC is long-term contracts with coal suppliers in nations with stable governments. Their interest does not stop at the purchase of coal but extends to the transportation and shipment. To be assured of a smooth flow, ATIC will obtain a participating interest in barge companies and in coal export terminals.

FRENCH AND AMERICAN COOPERATION

On November 1, the President of the French National Manufacturers' Association, Mr. Francois Ceyrack, accompanied by the Assistant to the President of France and the French Minister for Commercial Affairs in the United States, visited the port to determine steps necessary to expedite export of coal to France and imports of manufactured products to the United States. Additionally, ATIC has been in active discussion with the Port of New Orleans since 1979 on the possibilities of constructing an export coal facility on the lower Mississippi River. ATIC's current plans envision a joint venture of such a terminal with American principals. I, personally, expect some specific announcement by the French in the very near future as to the details of their joint venture. An early decision on this matter has been enhanced by President Reagan's support of a special effort to support fast-track legislation in congress for deepening the Mississippi River from 40 to 55 feet. Congressional leaders from several coastal states have introduced legislation to provide, on a fast-track basis, deeper access to the nation's ports, with special emphasis on New Orleans, Norfolk, and Mobile.

RAILROAD PORT INVESTMENT

A participating interest by the public and private sectors does not appear possible where the terminals are owned by the railroads and are served exclusively by rail. Terminals on the lower Mississippi River are capable of being served by both barge and rail. The Illinois Central Gulf Railroad is investing heavily in improvements of its trackage to New Orleans in anticipation of unit train movements of coal to and from the Illinois coalfields. At the board meeting of January 16, the ICG Vice President announced that with agreement with Ryan Walsh to operate the Port's bulk terminal, the railroad was placing an additional order for 400 coal cars. This competition with the barge lines will preclude increases in transportation costs such as have occurred in western areas where only one mode of transportation was available.

LARGE BULK CARRIERS AND PORT DEPTH

No port on the Atlantic or Gulf Coast has adequate channel depth to accommodate today's large bulk carriers. This is in contrast to ports such as Antwerp, with a 61-foot draft; Kashima, Japan, with a 72-foot draft; the Port of Fos, France, with a 55-foot draft; and the Port of Le Havre, France, with a 50-foot draft. Today, the workhorse of the bulk carrier trade is the 150,000 DWT vessel, requiring a 55-foot draft. In 1970, bulk carriers of only 100,000 DWT represented only 3% of the international bulk trade. By 1977, this had increased to 25% and it is increasing every year. American exports of grain and coal are penalized because these vessels cannot be fully loaded. Transportation savings on a ton of cargo from New Orleans to Europe, moving in a vessel loaded to a 50-foot draft versus the restriction imposed by the presently authorized 40-foot channel, is estimated at between $6 and $7. This could easily represent a billion dollars additional income to the producers by 1990.

For these reasons, the deepwater Louisiana ports on the lower Mississippi River, which serve a third of the continental United States, must be provided with an adequate access to the high seas. The Port of New Orleans not only strongly supports this project, but also supports all efforts to expedite the review, approval, and funding of the project. Representatives of the 18 mid-America states which make up the hinterland of this lower Mississippi River will also reap large economical benefits from the accomplishment of this project. If this project is not accomplished or expedited, the U.S. may lose much of the new trade opportunities to foreign competitors in world markets.

COST RECOVERY PLANS

In reference to a position on increased and/or new waterway user charges plus cost-sharing and cost-recovery proposals, we have very mixed feelings. On the one hand, we are opposed to them in principle because we feel that they are inflationary in that they are passed on to the consumer and that they also raise the cost of American exports in world markets. Inasmuch as the Administration and the Congress appear determined to move forward with some version of these charges this year, then we advise taking some insurance. We would recommend that fast-track deepening legislation incorporate not over 25 to 50% cost-recovery requirement with the final amount determined by the outcome of a six-month study to be carried out jointly by the Assistant Secretary of Commerce for Maritime Affairs and the U.S. Trade Representative to determine the impact of these charges on the competitiveness of the U.S. in international trade and the impact on the traditional distribution of U.S. commerce between ports and competing transportation modes. With regard to operation and maintenance charges, we would recommend the same 25 to 50% limitation pending the outcome of the previously recommended study.

We would like to make a number of specific recommendations on provisions for inclusion in any fast-tract deepening legislation that the Congress may pass. These are contained in Appendix C (page 187). Also, see Appendix D (page 189), a brief consultant's report covering the primary benefits to the petroleum industry from a 55-foot channel deepening of the lower Mississippi River from Baton Rouge to the Gulf. This report shows that there will also be substantial benefits for petroleum movements on the river once it is deepened. In addition, the reader is referred to an article from the April, 1981, issue of Coal Age that shows that Illinois coal can be delivered to Europe via the Gulf Coast at an $8-11 per ton saving over movement via the East Coast.

CHALLENGES OF COAL TERMINAL OPERATION
GERALD D. CUNNINGHAM

It is quite evident that if we are to take full advantage of
the opportunities that exporting Illinois Basin high sulfur coal
present, all the players will need to work together to form a
continuous chain of transportation and marketing logistics from
mine-mouth to end-user. While the Illinois Basin unfortunately
contains more higher sulfur coal than most of us would like, it is
fortunate to be located in one of the best transportation corridors
in the United States. The location of the more than 215 billion
tons of identified reserves is ideally suited for easy access to the
export market via the Mississippi River and the Port of New Orleans.
This coal can use the same rail and water transportation network
that has been in place and used for many years by the nation's grain
trade.

INLAND AND SEA LOGISTICS

New Orleans, situated at the lower end of the Mississippi River
near the Gulf of Mexico, has more direct connections to various
types of inland transportation than any other port in the United
States. Most ports are connected either by rail or water to inland
systems, but not by both. The East Coast ports of Hampton Roads and
Baltimore receive all of their coal by rail and have no access to
inland water transportation. This also applies to the West Coast
ports and to a lesser extent to the port of Mobile. Mobile does
receive small quantities of coal by the Alabama River from local
Alabama coalfields, but until it is connected to the Tennessee and
Tombigbee Rivers in the late 1980s, Mobile will not have easy access
to the U.S. water systems. New Orleans, however, via the
Mississippi River, has direct access to most of the major U.S.
rivers. This opens up the Midwest coal areas, the Illinois Basin,
the Arkansas Basin, and the east Kentucky and Tennessee. This river
system provides for these regions barge transportation down to the
port of New Orleans for export via the Gulf of Mexico.

New Orleans also links into the U.S. rail system via the
Illinois Central Railroad, TPMP, and Southern, and South Pacific
Railroads that terminate at New Orleans and service a large part of
the U.S. coal-producing areas. The East Coast ports are served by

rail but only from the eastern coalfields of West Virginia, Virginia, Pennsylvania, and east Kentucky. The primary railroads serving these ports are the N & W, B & O, and C & O. The West Coast is not yet a major factor. The only access to the western coalfields is by rail. These connections, as well as adequate port facilities, have to be developed before large movements can begin from the western areas. Overall, New Orleans is better situated than either the East Coast or West Coast ports to take advantage of the various inland transportation systems.

EAST COAST VS. GULF EXPORTS

The economics of exporting coal through the port of New Orleans, as compared with the East Coast and other Gulf Coast ports, is a function of the mine origin. Studies at International Matex have concluded that coal destined for export, produced east of an imaginary line from west Pennsylvania running south through West Virginia and eastern Kentucky, is more economically shipped via the East Coast terminals compared to the Gulf Coast and New Orleans. This conclusion does not give consideration to demurrage costs for either rail or vessels, but is supported by all rail rates from these areas to the East Coast ports compared to any integrated rail-water rate to New Orleans or Mobile. However, beyond this area, the East Coast ports are generally not economically competitive with New Orleans.

The port of Mobile located on the Gulf Coast has economic advantages over New Orleans only when the coal comes from Alabama or east Tennessee origins. Mobile could be more of a factor if and when the Tenn-Tom Waterway is completed, although the new connection limits barge tow sizes. The other major port located on the Gulf Coast is Houston. There are several plans to build coal terminals in that area (Pelican Island and others). This port could be an export terminal for western coals because of rail connections with those producing areas, but it is doubtful that the railroad can produce lower unit train rates in the near future to compete with rail/barge movements to New Orleans. Most western coals in the future will probably move to the West Coast for export through new terminals to the Far East.

By combining both rail and water transportation, New Orleans has definite economic advantages over the other ports for coal origins in the Illinois Basin and Arkansas and some western areas, since a New Orleans coal terminal has the flexibility of integrating rail and barge.

While New Orleans compares favorably to other ports for inland transportation, it does not, however, have that advantage with ocean rates. Our latest information suggests that coal exported from New Orleans and bound for Western Europe is $2 to $4 higher for ocean freight than the East Coast rates when shipped in 60,000 DWT vessels. This differential will be reduced when larger vessels can be used following the dredging of the Mississippi River to 50 or 55 feet. However, present ship demurrage rates at the East Coast ports are in excess of the ocean freight differentials. Therefore, the challenge is to assure the Illinois Basin producers as well as foreign buyers that sufficient port capacity at reasonable cost will be available to complete the chain that is necessary for the successful marketing of high sulfur coal in the world market. To date, the problem has been a lack of new and modern terminal capacity at the Gulf Coast. However, private industry has accepted the challenge during the past year with the announcement of several new coal export terminals to be constructed, as well as the upgrading of the two existing terminals located below New Orleans in Plaquemines Parish. To this new capacity we must add the substantial interim capacity of midstream loading that has materialized in recent months. A summary of existing and planned export capacity is presented in Table 1.

Table 1
EXPORT CAPACITY

EXISTING:

NAME	CAPACITY	EXPANSION PLANNED
Electro Coal	10 MTPY	20 MTPY
International Marine Terminal	4 MTPY	12 MTPY
Burnside Terminals	4 MTPY	-------

PLANNED:

International Matex Terminal	12/15 MTPY	20 MTPY
Ohio Barge Line	11 MTPY	20 MTPY
Peabody Coal	10 MTPY	-------
Freeport Sulfur	-------	-------

Many other firms have shown an interest in either building or participating in a Mississippi River coal terminal. Assuming some of these planned facilities go no further than the planning stages, it is apparent that sufficient capacity will be on-stream between now and 1983 to serve the most optimistic projections of coal that would be exported via the Gulf coast.

To make New Orleans a more attractive coal port, we must have immediate action to dredge the southwest pass at the mouth of the Mississippi River to a depth of 50 to 55 feet, which will be essential to accommodate the larger and more economical bulk-carrying ships in the 100,000 to 150,000 DWT class. The economics to be gained from shipping coal in these larger ships ranges from $5-$8 per ton. Therefore, it is imperative that this project receive the support of all of us if we are to remain competitive with other countries that already have deep draft ports.

International Matex recently announced the construction of a major new coal export terminal to be located at mile-point 46-47 on the Mississippi River about 40 miles below New Orleans. This site is known as the Magnolia Plantation and comprises over 2,000 acres of land with over 2 miles of deep Mississippi River frontage. This gives Matex the ability to fleet about 180 barges and to handle 15 to 30 large tows intact. Current plans call for the construction of a dock designed to handle ships to 130,000 DWT and a traveling shiploader with a design loading rate of 6,800 tons per hour. A continuous barge unloader with a design rate of 5,000 tons per hour will also be incorporated with a barge cleanout station. By using two-track mounted stacker/reclaimers and additional stackers, the terminal will have the full flexibility to load ships while unloading river barges at the same time. Phase I will provide about 850,000 tons of live storage with an additional 1 million tons of dead storage. Services provided by the terminal will include the mixing of various types of coal to the customers' specifications while reclaiming from two different piles at the same time. An automatic sampling system is also being designed into the system for quality control. International Matex believes that a terminal should be designed and operated to offer the customer the greatest amount of flexibility.

There are certain benefits accruing to an independent public terminal not controlled by any coal producers or buyers. It will enable more coal producers to participate in the export market without making the large capital commitment needed for their own terminal, and, at the same time, guarantee their throughput needs without preferred treatment of the terminal operator's product for

loading during peak times. A customer pays only for his capacity
and can trade with others in the terminal as oil companies have done
for years. The assets are shared by many companies, while giving
each the flexibility needed on an equal basis.

VI

International Coal Export Policy
Implications: The U.S. Perspective

INTERNATIONAL ECONOMIC POLICY IMPLICATIONS
DONALD V. EARNSHAW

The United States, by expanding as a major international source
for steam coal, can significantly reduce the world's dependence on
petroleum for its energy needs. It would be natural for countries
seeking a stable energy source for commercial and political reasons
to choose the United States as a major supplier. Long-term
commitments or contracts between United States producers and foreign
buyers provide the assurance each party needs to obtain the capital
for, and to initiate required improvements or investments in, new
equipment for production, handling, and movement of the coal. The
foreign buyer gains additional security against economic fluc-
tuations and an assured source of supply that meets established
quality standards. The United States only needs to reinforce its
reputation as a reliable supplier of coal and to establish a
definitive policy that the government will not interrupt contractual
arrangements unless required to do so by a national emergency. The
United States has never restricted coal exports, even during strikes
that have reduced domestic supplies. This policy for commercial
stability in coal trade also extends to allowing foreign investment
in coal projects, consistent with existing federal and state laws.
Such a policy not only provides needed capital, but further binds
foreign purchasers to long-term commitments to buy American coal.

MAINTAINING A STRONG DOMESTIC INDUSTRIAL SECTOR

In addition to trade balance benefits, substantial increases in
steam coal exports could provide impetus in areas of industrial
activity long neglected, specifically the following:
1.   the rebuilding of our deteriorating railroad beds;
2.   increasing our rolling stock and expanding the handling
     capabilities and efficiencies of our ports;
3.   encouraging the repair and improvement of our highway systems
     serving our ports; and
4.   fostering improvements to our inland waterway system.

HIGH SULFUR COAL EXPORTS

At present, high sulfur coal finds markets in the industrial use side, such as cement and lime kilns, rather than in power generation. Significant controls on combustion emissions resulting from coal use are imposed by most coal importing nations which aim for a sulfur content of 1.5% or less. A major influence on coal quality purchasing decisions is the historic pattern of consumption. In Europe, for example, most domestic reserves are good quality, low sulfur bituminous coals. As use of imported coal becomes more prevalent, customers are expected to seek comparable coal properties. While near-term U.S. coal exports may be low in sulfur, new technologies are being researched and developed that will enhance the future sale of high sulfur coal. Such technologies include flue gas desulfurization, fluidized bed coal combustion, and synthetics. Thus, long-term prospects appear good not only for the export of U.S. high sulfur coal but also for the technologies and accompanying equipment being developed in the United States to utilize all qualities of coal. Estimates place U.S. coal exports to be one-quarter bituminous with more than 1% sulfur by mid-1990. Bituminous coal with less than 1% sulfur and low sulfur subbituminous coal will account for three-quarters of total coal exports.

DEPARTMENT OF COMMERCE ACTIVITIES

Many programs exist within the International Trade Administration of the Department of Commerce to promote coal exports although they are not presently devoted exclusively to coal exports. The Department of Commerce is currently examining how these existing programs and planned initiatives can be directed within current resources for special emphasis on coal exports. Mechanisms for coordinating these activities with those of other organizations with some coal promotion responsibilities are also being looked at within the department. The Department of Commerce took an active part in the recent Interagency Coal Export Task Force. The International Trade Administration of the Department of Commerce has continued its efforts to assist the U.S. coal industry to improve its export capability, to aid foreign buyers to locate suitable United States producers and U.S. manufacturers of coal-related equipment and technology, and to identify potential purchasers for American exporters.

The International Trade Administration has some 40 programs and activities designed to assist exporters at all levels of potential and capability. Many of these programs are particularly suited to assist in the promotion of the export of United States coal. Specific plans have been formulated to cooperate with U.S. industry in carrying out an extended program to promote coal exports.

COAL VS. OIL: U.S. RESEARCH INTERESTS
ROGER W. A. LeGASSIE

There is no question that coal must assume growing importance in the world's energy profile in the coming decade and beyond. It is a uniform viewpoint that the amount of oil in international trade probably will not increase significantly from 1980 levels, especially given increasing oil demands within OPEC and other countries with emerging oil production potential.

On the other hand, coal has more potential than any other fuel source for providing additional energy in the 1980s. One reason is economics. Given the recent multiple increases in world oil prices, coal has become a significantly cheaper fuel than oil for utility and large industrial boilers, both domestically and internationally. To illustrate this, let us assume a world oil price of $35 per barrel and a delivered coal price of $60 per short ton. The total annualized costs for a new coal-fired, baseload powerplant in Europe can be calculated at approximately 50.38 mils per kilowatt-hour--a figure which includes the costs and capacity penalty of adding pollution control equipment such as a scrubber. The costs of fuel and operations alone for an existing oil-fired powerplant would be approximately 57.50 mils per kilowatt-hour.

In other words, many countries may find it more economical to build new coal-fired powerplants than to continue operating existing oil-fired plants. This economic advantage is providing the impetus for greatly expanded world steam coal trade, and it is important to note that coal retains these benefits even after adding the cost of pollution control equipment.

Another contributing factor is, of course, the magnitude of the nation's storehouse of coal. The Department of Energy in May 1981 released revised official estimates of our domestic coal reserves as of January 1, 1979. The nation's coal reserves were estimated at 475 billion tons--up 8% from the 438 billion tons estimated in 1976. This is the coal that can be economically recovered by current mining techniques. The coal reserves beneath the U.S. are equivalent, in terms of energy content, to almost 2 trillion barrels of oil. Bituminous coal is the principal rank of coal in the reserve base, accounting for 242 billion tons, or about 51% of the total.

In light of these factors, the contribution of U.S. coal in meeting an expanding world energy demand will continue to increase. Last year, steam coal exports jumped 90% over the 1979 level.

Exports to Europe increased by 700%. The coal export forecasts for
1985 by the World Coal Study were reached last year, five years
ahead of the projections. Between now and 1990, the Interagency
Coal Export Task Force has projected that the U.S. portion of
international trade in steam coal could increase to approximately 64
million tons per year.

COAL AND INTERNATIONAL REGULATIONS

Yet, as bright as coal's future appears based on these
hypothetical projections, one must be cautious in drawing
conclusions without first examining several factors more
specifically. In particular, when estimating the potential for high
sulfur coal exports, it is important to be aware of the limitations
imposed by potential steam coal buyers on undesirable emissions.
Most coal-importing countries generally impose some degree of
controls on the levels of sulfur dioxide and trioxide emissions that
can be emitted by coal combustion.

Controls on sulfur emissions take many forms: limits on fuel
sulfur content; limits on emissions from a single plant, or on
overall nationwide emissions; and tall stack or other design and
equipment requirements. Regulatory structures vary from country to
country. In some cases, the setting of air quality standards and
their enforcement are shared by local jurisdictions and the national
governments; in others, it is totally centralized. Some countries
have created "protected zones," in which utilities and industries
agree to adhere to stricter standards, either at all times or during
pollution alerts.

Italy, for example, limits utility coal sulfur content to 1.0%,
although exceptions have been granted for power plants burning
domestic lignite. The Netherlands has an emissions ceiling of
500,000 tons of sulfur dioxide per year from all sources. Although
use of electrostatic precipitators to reduce the release of
particulates is common, application of flue gas desulfurization
equipment is not widespread. Only in Japan, where flue gas cleanup
systems are required for nearly all power plants, has such
technology been widely applied. Consequently, European utilities
are expected to seek coals that are relatively low in sulfur
content, generally 1.5% or less. Tighter restrictions will apply in
specific areas. Similar requirements are expected in the Pacific
Rim.

Higher sulfur levels are generally acceptable for use in cement
and lime kilns in both Western Europe and the Far East. In some
countries, restrictions are less stringent for industrial uses than
for utilities, and higher sulfur coals can be used. The coal demand

for these uses will, of course, be substantially less than for power generation.

For these reasons, estimates are that nearly half the coal exported from the U.S. in the mid 1990s will most likely be bituminous coal of less than 1% sulfur. The rest will be almost equally divided between low sulfur subbituminous coal and bituminous coal of greater than 1% sulfur.

EMERGING U.S. TECHNOLOGIES

While the major proportion of coal exports from the U.S. in the next several years is expected to be low sulfur coal, the longer-term prospects could be enhanced by the application of new, more environmentally beneficial technologies. One such example is the fluidized bed coal combustor.

A fluidized bed combustor burns coal in a turbulent bed of hot limestone or dolomite, suspended by an upward-blowing stream of air. The limestone or dolomite chemically combines with the sulfur dioxide formed during the combustion process and prevents it from exiting the boiler. No scrubber is required to burn high sulfur coal. In the U.S., fluidized bed technology on an industrial scale is approaching the commercial threshold. Several firms are moving into commercial application of the atmospheric pressure version of this advanced technology, based in large part on the successful demonstration conducted at Georgetown University in Washington, D.C., and a sound technical data base developed in both private and public sectors.

FLUIDIZED BED COMBUSTION

Fluidized bed coal combustion has attracted similar interest internationally. In the Federal Republic of Germany, for example, fluidized bed combustion represents a significant area of deve-lopment in a government budget where about half of the non-nuclear research funds are devoted to coal. Two atmospheric fluidized bed boilers--the largest with a thermal capacity of 35 Mw--are currently being operated by Ruhrkohle in cooperation with several private firms, and a 124 Mw plant is planned for construction in Hameln by a local utility. In the United Kingdom, the British Babcock atmo-spheric boiler at Renfrew has accumulated substantial operating hours, and a commercial expression of this is the retrofit instal-lation of a fluidized bed boiler by Babcock International Combustion at the Central Ohio Psychiatric Hospital in Columbus. There are at least nine companies in the U.K. which are prepared to offer on a commercial basis a fluidized bed boiler or furnace. In Denmark, where 99% of the country's energy is imported, small-scale 5-10 Mw fluidized bed boilers are being studied for potential use in district heating systems. In Japan, where more than 15 million tons of coal will be used in power plants in 1990, fluidized bed technology is gaining increasing attention, and the Japanese have a 5 Mw fluidized bed pilot plant under construction. Additional coal will be imported for industrial applications.

The most widespread use of fluidized bed combustion is in the People's Republic of China, which revealed at last year's Sixth International Conference on Fluidized Bed Combustion that it had more than 2,000 fluidized bed boilers in operation, some with more than 40,000 hours of accumulated experience.

Another longer-range technology area is synthetic fuels--oil and gas manufactured from coal. Here too, considerable international activity has become apparent. For more than 20 years following World War II, the largest commercial synthetic fuels activity in the world was centered in the Republic of South Africa which last year activated its second commercial coal-to-liquids facility. In other countries, several smaller-scale synthetic fuel systems are also operating predominantly to produce coal gas for use in manufacturing chemicals. Like the South African plants, many use technology suited for the coals found in their general region, typically low sulfur, non-caking coals. The processes encounter considerable difficulty when applied to the high sulfur bituminous coals found in the U.S., due to the tendency of these U.S. coals to cake under the required temperatures and pressures.

SYNTHETIC FUELS IN THE U.S.

More recently, the turbulent nature of the world oil situation has sparked renewed interest in the commercial development of new, more efficient synthetic fuels technologies in several countries, some without indigenous coal reserves. In many cases, U.S. technology, including that which can process high-sulfur bituminous coals, is being pursued with the eventual expectation of importing U.S. coal as a primary feedstock.

The Exxon Corporation, for example, proposes to build a $350-500 million coal gasification pilot plant in the Netherlands as a precursor to commercial deployment of the technology in Europe. Several coals will be tested including U.S. high sulfur caking coals. The Texaco coal gasification process, also suited for high sulfur caking coals, has been licensed to the German firms of Ruhrkohle and Ruhrchemie, and they have continued development of the concept in a 150-ton-per-day pilot plant at Oberhausen.

International Shell is developing a pressurized version of the Koppers-Totzek gasifier and is now operating a pilot plant near Hamburg, West Germany. The British Gas Corporation plans to continue development of an improved version of the first-generation Lurgi gasifier, which has as one of its benefits the capability to use midwestern and eastern U.S. coals.

Future markets for these gasifiers could be enhanced by the application of new technologies such as the M-gasoline process. This technique is attracting increasing attention in Europe, as well as in the U.S., as a way of converting methanol (which can be made from coal gas) into high-octane gasoline.

In the U.S., Ruhrkohle of West Germany, a Japanese consortium called Japan Coal Liquefaction Development Co., and AGIP of Italy are co-sponsors of the Exxon Donor Solvent Process--an advanced coal liquefaction method now at the pilot plant stage. Ruhrkohle is also participating in the H-coal process development, another advanced coal liquefaction pilot plant.

THE FEDERAL ROLE IN COAL INVESTMENTS

With increasing private sector interest, both in the U.S. and many other industrialized nations, coupled with the need to restore the nation's economic vitality, the Administration has made several fundamental changes in federal policies regarding the development of our energy resources. Simply stated, we see the principal near-term federal role as improving the investment climate, providing a competitive marketplace and implementing tax and regulatory reforms. Decontrol of oil prices was the first step in this direction. As oil is priced at its replacement value, coal will become an even more attractive alternative, and by improving the investment climate, industry will be encouraged to increase its R & D and capital investment spending. In addition, we are reviewing for withdrawal more than 200 regulations issued by DOE or its predecessor agencies.

The Clean Air Act comes before Congress for review this year. In this regard, we are examining ways in which a better balance can be brought between our nation's need for energy production and our need for a safe and healthy environment. The Act, as it has been interpreted, has put the federal government in the position of dominating decision making related to industrial development and growth. This role has traditionally been exercised by the states and the private sector. We anticipate that formal Administration recommendations will be sent to the Congress. Recognition of the need to channel capital investments toward the environmental

problems and solutions that offer the greatest payoffs will be emphasized in these recommendations.

REGULATIONS AND TAX REFORM

The Surface Mining Control and Reclamation Act is also being
reviewed. The Act established a complex body of laws to regulate
all surface mining activities and the effects of underground mining
on the surface environment. In concert with the Vice President's
Task Force on Regulatory Relief, the Department of Interior is
reconsidering the regulations and intends to remove burdensome,
excessive, and counterproductive requirements.

In the area of tax reform, the Administration's proposed
Accelerated Cost Recovery System would stimulate plant and equipment
expenditures for new coal facilities by increasing the after-tax
rate of return for these investments and by increasing a firm's
after-tax cash flow. The proposal provides accelerated write-off of
capital costs of machinery and equipment and certain industrial and
commercial buildings over a period of 10, 5, or 3 years. Coal
mining equipment which is generally depreciable over an 8-12-year
period under current law, could be written off in 5 years under this
new system.

Regarding coal exports specifically, the Department of Energy
was one of 14 departments and agencies which participated in the
Interagency Coal Export Task Force formed in the spring of 1980 at
the direction of the past Administration. In its roughly 6 months
of existence, the Task Force undertook a number of related
activities including the assembly of existing data and the
development of significant new information regarding the
international coal market. In January 1981, the Task Force
published an interim report for public comment. To date, more than
40 comments have been received from a broad spectrum of interested
groups and individuals. Many of the comments stressed the opinion
that exporting coal is primarily a private sector responsibility and
that the proper role of government is to ensure that there are no
federally imposed barriers to more efficient delivery of coal to
ports and waiting vessels, and that the federal permitting and
review process be streamlined to prevent inordinate delay in port
improvement projects.

COAL EXPORTS POLICY

The Administration is currently reviewing the comments prior to
determining what, if any, specific federal measures are needed as a

follow-up to the draft Interagency report. This activity is receiving Cabinet-level attention, and we anticipate the Administration will be prepared to comment specifically on the subject of coal exports within a short time.

We have taken action recently within the Department of Energy to establish a specific function related to coal exports within the Office of Fossil Energy. Mr. Mario Cardullo will serve as the liaison within DOE for these activities.

Recognizing that specific comments are still forthcoming from the Administration, I have attempted to summarize some key factors impacting the U.S. position in international coal trade.

Port-handling facilities are being expanded. As demand increases, market forces have provided incentives to industry to do what is necessary to promote exports. Today, much is already being done through private initiative. Expansion projects are underway or planned at some 35 ports and harbors. As a result of numerous short-term efforts, coal exports in March 1981 (9.6 million short tons), were almost 42% over February, and 72.4% over March 1980. Continuation of this positive trend is dependent, of course, on an early resolution of the coal strike.

THE NEED FOR PORT EXPANSION

Existing effective loading capacity is estimated at 94 million short tons per year. Expansion underway would add more than 23 million tons by 1983 and planned expansion would increase capacity to 184 million tons. Given current projections for U.S. coal exports, port capacity should be more than adequate within a few years.

After 1990 a significant portion of coal will be transported in ships too large to call at present U.S. ports unless alternate coal loading technologies are developed and available (such as single-point mooring for either slurry or coarse coal loading). Unfortunately, primarily due to the existing review, authorization, appropriation, and permitting process, past port and navigation improvements have taken 20 to 25 years from the initial study to final construction.

The Department of Energy supports efforts to speed up the review, permitting, and authorization processes associated with port projects. The Administration has proposed legislation providing private sector reimbursement in the form of user fees for costs associated with port dredging. This will help port improvement projects proceed at a pace consistent with commercial demand rather than federal budgetary constraints.

THE DOE TECHNOLOGY PROGRAM

By turning toward a free, competitive marketplace and away from massive government subsidies, private industry will be able to do what it does best--namely invest funds on those activities which improve a firm's competitive position because they are cost-effective. In turn, the federal government will focus its funding support on that longer-term, high-risk research and development which industry has generally been less willing to undertake.

A prime example of this refocused approach and its specific application in the coal area is in our environmental control technology program. Industry is making significant progress in the commercialization of advanced scrubber systems, such as the lime spray dryer. With several commercial units being marketed for low sulfur western coal, our original FY 1981 program was to accelerate the application with eastern high sulfur coals. This plan was altered when it became evident that the momentum building in the private sector was beginning to penetrate the eastern coal industrial market and a decision was made to rely on this momentum to carry into the utility sector without major governmental assistance.

Likewise, we are concluding our development efforts for the atmospheric version of the fluidized bed combustor which is gaining increasing domestic and international commercial interest. We are now focusing our program on the next generations of the technology, including the pressurized fluidized bed as well as new concepts such as staged combustion. We are also beginning to examine new coal-based fuels, such as coal-water, coal-methanol and coal-ethanol mixtures, in view of the activity underway both domestically and overseas in commercially applying coal-oil mixture preparation and utilization technology.

In the synthetic fuels area, the Department of Energy is ending its program of major technical fossil energy demonstration, and, instead will focus solely on supporting longer-range research and development. The responsibility for constructing major synthetic fuels facilities, both commercial-scale and "first-module" demonstrations, will be consolidated within the U.S. Synthetic Fuels Corporation. This will provide a more efficient and focused program for demonstrating synthetic fuel production with a concomitant savings of federal dollars.

Our coal mining and preparation program has also been refocused to emphasize long-range/high-payoff/high-risk research. No longer will we fund activities that are designed to improve a specific company's process; rather we will focus on generic and more basic R & D which can benefit the industry as a whole. Underground mining research is concentrating on the following: 1) the development of basic mining technologies including strata control, seam and rock strata cutting, materials handling, automation and system control, and environmental monitoring and control; 2) the development of advanced, innovative concepts of mining underutilized reserves such as coal seams which are too thin for conventional mining systems or which lie at too steep an angle; and 3) integrated coal mining/preparation concepts such as underground coal preparation.

MINING RESEARCH IN THE U.S.

DOE has not requested funding for surface mining research in the fiscal 1982 budget proposal now before the Congress. Since many of the near-term projects are initiated by private companies and since the improvements in productivity would accrue to the individual companies, industry has sufficient incentives to pursue this kind of lower risk technology development without continued federal funding.

Coal preparation research will focus on the development of 1) a comprehensive data base on the potential for beneficiating principal U.S. coals; 2) technologies that can extract 85-95% of the total sulfur and ash from these coals; and 3) environmentally-oriented research such as the characterization of coal preparation plant wastes for possible disposal in underground mines.

To coordinate our mining research, we have implemented memoranda of understanding with the two other federal agencies involved in the mining area--the Departments of Interior (DOI) and Labor (DOL). With DOI's Bureau of Mines, which is charged with coal mine health and safety research, DOE reviews its coal mining production research to avoid duplication and, maximize complementary research efforts. With DOL's Mine Safety and Health Administration (MSHA), which is charged with enforcement of federal coal mining health and safety regulations and approval for safety of all equipment and methods employed in U.S. mines, we review our research efforts, alert MSHA regarding needs for advanced mining system approval, and solicit suggestions for improving the technology under development in our program.

To ensure that our international partners are aware of the technology under development in the U.S., the Department of Energy actively participates as one of 20 countries in the International Energy Agency, including its Economic Assessment Service. This Service assesses the economic merits of existing and emerging technologies and provides nations with key data for their use in determining the commercial potential within their own economic sphere.

An integral part of our program is the conviction that we must be as concerned about the energy security of our allies overseas as we are about our own vulnerability to energy disruptions and our reliance on unreliable energy sources. We can make positive contributions in both areas by reducing our own oil imports, thereby relieving price pressures on the world market, and by becoming a major exporter of steam coal.

These are not overnight tasks, and they will not be accomplished with massive government spending and Washington-based decision-making, but by the private sector. Our task in government is to ensure that we do not stifle the initiative and technologic creativity that exists within our private enterprise system to use U.S. coal in the international marketplace. It is more properly our task to ensure a free and stable coal trading environment.

INTERNATIONAL PERSPECTIVE
JOHN P. FERRITER

The Administration strongly supports expanded coal exports and is committed to a course of action to facilitate coal sales abroad. We are convinced that increased coal exports will contribute to our national interest in a number of ways. First, they bolster our economy by strengthening our balance of payments and expanding employment in key regions of the nation. Second, they contribute to our energy goals by reducing world competition for petroleum. Third, they enhance our national security by reducing our allies' dependence on unreliable sources of energy. It is this third attribute of expanded coal exports that relates most directly to the prospects for "emergency" utilization" which I will address today.

Since the oil crisis of 1973-74, we and our allies have endeavored to improve our energy security. Through coordinated national and international efforts, we have sought to ensure that precipitous oil supply interruptions would not cause an "energy emergency." Broadly stated, the following have been our immediate goals:

1. to diversify our energy mix;
2. to reduce our dependence on imported oil; and
3. to develop alternative sources of energy which could be tapped in a crisis.

COAL AS A REPLACEMENT FOR OIL

We have made steady progress toward these objectives, but a great deal remains to be accomplished. The 21 industrialized member-countries of the International Energy Agency (IEA) still depend on oil for almost 60% of their energy needs. This dependence is not expected to fall below 50% until after 1990. Furthermore, IEA members import 55% of their total oil consumption. Most countries do not yet have sufficient strategic energy reserves to cope with a supply interruption. As events of the past several years have demonstrated, interruptions of even a small portion of oil imports can have very serious consequences for the economies of all oil-consuming nations.

We believe, and our allies concur, that use of coal must grow dramatically as our economies continue to shift from oil to other energy resources. First, coal provides an economical alternative. In recent years, the prices of oil and natural gas have risen so rapidly that those fuels now cost two to three times as much as coal for an equivalent amount of energy. Even mined underground, transported long distances over land and sea, and burned in plants fully equipped for environmental protection, coal is, and is generally expected to remain, substantially cheaper than oil on a dollar-per-Btu-generated basis.

COAL AS A SECURE RESOURCE

Second, and more directly relevant to our discussion, coal is particularly well-suited to enhance a country's energy security. IEA countries have all committed themselves to diversification of the fuels upon which they depend. Use of coal as a substitute for oil in industrial furnaces and power plants will contribute to the diversity and reduced concentration of national energy mixes, the countries from which energy resources are purchased, and the types of fuels imported. Supplies of coal from secure sources will be plentiful for the foreseeable future. Furthermore, oil- and gas-fueled furnaces can be built with a dual-fire capability, so that coal can be substituted for other fuels without significant capital investment or shutdown time.

COAL AND NATURAL GAS IN EUROPE

Coal's potential contribution to energy security, particularly in the case of European countries, is even more impressive if we compare it to natural gas, the other most readily available substitute for oil. Europe already imports oil from the most likely prospective suppliers of natural gas: the Soviet Union, Nigeria, Libya, and Algeria. Proposed gas imports from these countries could more than offset reduced oil purchases from these same countries. Therefore, increased European gas imports might reinforce, rather than reduce, European dependence on specific petroleum suppliers.

This development would be of most concern with respect to potential European energy dependence on the Soviet Union. Six key European countries, West Germany, France, Italy, Belgium, Austria, and the Netherlands, are currently negotiating a massive natural gas deal with the Soviet Union. If they agree upon a long-term supply relationship and decide to build the proposed Western Siberia to Western Europe Pipeline, by 1990, the Soviet Union could provide as much as 20% of all Western European gas imports. This would represent 5% of their total energy consumption. In individual countries, dependence could run even higher. In West Germany, for example, Soviet gas could represent up to 30% of total gas consumption. Such significant dependence on Soviet energy supplies might have serious political ramifications.

It is clear that if the Europeans decide to import large amounts of Soviet gas that they will have to take precautions to ensure that this commercial tie does not foster unacceptable economic or political vulnerability. In particular, Europeans might seek to increase their use of coal from the U.S. as a means to counterbalance partially their increased use of Soviet gas.

The energy/security considerations for coal use in Japan and less developed countries are not as significant. The Japanese already purchase oil and gas from a diverse pool of suppliers, and the Soviet Union is not looking to export large amounts of petroleum to Japan or nonaligned countries. Yet in these cases, the economic motivations for increased coal consumption still hold.

It appears, then, that over the next decade, international demand for coal, based on commercial and security considerations, could grow significantly. Conversely, expanded U.S. coal exports could make a great contribution to our nation's economic welfare and national security.

RELIABILITY AND LONG-TERM COAL CONTRACTS

There are, nonetheless, some issues remaining to be resolved, which will in large part determine the distribution of market shares of the expanded international coal trade. The key issue on both the demand and supply sides is captured in a word--reliability.

Foreign consumers are currently reluctant to enter into long-term contracts. Bottlenecks caused by insufficient coal infrastructures presently limit U.S. exports. In the last year, political unrest in Poland and miners' strikes in Australia and the United States cut international coal shipments. Prospective buyers will be unwilling to make a long-term commitment to a particular supplier unless they are confident that they will be able to take delivery of coal on a timely and economical basis.

Of course, coal producers face a related problem of reliability. The coal industries may not increase their production capacity and upgrade the export infrastructure as quickly as possible until they receive assurances of long-term demand that would justify the requisite investments.

This type of chicken-and-egg problem based on reliability which could threaten expanded coal trade and its commercial and security contributions is apparently not retarding U.S. coal development and sales. The government has taken some steps to alleviate these problems. At the 1979 Tokyo Economic Summit, Western leaders pledged not to interrupt coal exports under long-term contracts, unless required to do so by a national emergency. At Venice in 1980, the leaders announced their agreement to encourage coal producers and consumers to agree to long-term contracts. We currently are seeking our allies' official sanction of foreign investment for the improvement of the U.S. coal export infrastructure.

U.S. leaders have assured foreigners that coal export contracts are enforceable in U.S. courts and pointed out that our country has never restricted coal exports, even during strikes that reduced domestic supplies. Furthermore, the Administration has already indicated its intention to minimize, consistent with national interests, the regulatory burden on coal and associated industries.

U.S. COAL AND DEVELOPING COUNTRIES

We are also trying to increase the use of coal in the developing world. Many developing countries have little experience with coal as a fuel and do not have the technical infrastructure necessary for coal use. Over the past several years, the U.S. Government, the World Bank, and others have offered technical assistance for coal-related projects in developing countries. Under the auspices of the AID Trade and Development Program, coal-related projects are underway in a number of countries.

PROBLEMS FOR SOLUTION BY THE PRIVATE SECTOR

Nonetheless, the government efforts are just a start. The task of selling U.S. coal abroad must be taken up by the private sector. I am pleased to note that industry is already rising to that challenge. Private companies are acting in all of the areas we might identify as helpful:

1. expanding port and inland transportation infrastructure;
2. expanding coal production;
3. financing both of the above;
4. seeking to lower costs;
5. seeking to enhance reliability in fulfillment of contract provisions (timing of delivery, prices, quality specifications, avoidance of disruptions in the flow of coal supplies, etc.); and
6. increasing responsiveness to purchase requirements of foreign markets and buyers.

There are plans to expand port facilities so that they will be large enough to handle projected coal shipments by 1985.

THE ROLE OF U.S. GOVERNMENT IN COAL EXPORTS

In our view, actions in all of the above areas should be governed by normal considerations of profitability and should therefore be left in the private domain. The U.S. Government will not subsidize coal exports, nor will it interpose itself in any other way directly in coal trade.

However, in light of the energy security issues related to coal trade, the government has an important role to play. We are, therefore, providing support to U.S. coal producers by emphasizing to our foreign interlocutors, both bilaterally and in the IEA, the need for them to provide assurances (e.g., through agreement to long-term coal purchase contracts or investment in U.S. coal production and transportation projects) of their long-term demand for U.S. coal. We believe that such assurances will encourage further expansion of the export infrastructure. Expansion of the infrastructure, in turn, is likely to stimulate further foreign interest in demand for coal.

In sum, we believe that the commercial demand and strategic importance of U.S. coal exports will expand during the current decade. The executive branch will continue to work with foreign buyers and governments to foster demand for U.S. coal and to enhance the reputation of the United States as a reliable and competitive coal supplier. At the same time, we will continue to depend on our strong American tradition of reliance on the private sector for the conduct of coal trade.

VII

Utilization of High Sulfur Coal

BACKGROUND AND SUMMARY
ROBERT JACKSON

The simplest way to use coal is to burn it directly. Because it is a solid, however, it can only burn in equipment that burns a solid. It is, therefore, quite unsuitable for burning in a motor car or a truck. When coal is burned, it produces two side effects: it leaves behind ash which has to be disposed of; and it puts into the atmosphere various quantities of contaminants contained in it. Disposal of the ash has been handled satisfactorily for many years, but today the venting of the combustion products of sulfur is no longer acceptable. This problem has been attacked, where required by the addition of scrubbers to boiler stacks, but a more important solution may be the use of fluidized bed boilers.

FLUIDIZED BED COMBUSTION

In these boilers, coal is burned in a bed of solids and fluidized by combustion air; limestone is added to react with the sulfur oxides as they are produced. Until now, the designs proposed have had some size limitations which do not make them very applicable to the utility industry--the largest user of coal. Now, however, considerable progress has been made in the development of circulating fluidized bed boilers which do not suffer from the above limitation.

Three boilers of this type have passed the pilot plant state of development. One of these, following a Battelle design and capable of producing 50,000 lb/hr of steam, is under construction by Struthers for Conoco as part of a heavy crude oil production in Texas. A second, being developed by Conoco Coal Development Company in association with Stone and Webster, has successfully completed pilot plant tests, and work has begun on a 50,000 lb/hr design for one of Conoco's plants. The third design is by Lurgi and is based on many years of experience on retorts and calciners. A pilot plant is being built and designs are being made for industrial and utility boilers.

LIQUEFACTION

All of these designs can be used at the size needed for utility applications. Their successful development could permit a considerable expansion of the market for high sulfur coal--although not until the end of this decade. The most attractive method for using coal, because the largest fuel markets are for liquids, is to separate the unwanted ash and sulfur as soon as the coal is mined and convert the coal into a liquid. The coal can be liquefied directly by dissolving it into a liquid to allow the nonsoluble ash to be separated by filtration from the soluble part and the solvent, or the coal can be indirectly liquefied by reacting it with steam and oxygen to produce a gas mixture which can be used as feed to a reactor in which liquid is synthesized.

Commercial interest in coal liquefaction developed in Germany with the invention of the Bergius Process in the 1930's. This process involves the reaction of coal with molecular hydrogen at high pressure. By 1943 the total installed hydrogenation capacity including coal and coal tar feedstocks amounted to slightly more than 4 million tons per year. The hydrogenation of coal and coal tar as a commercial industry ceased to exist after 1945 primarily because it was not competitive with natural petroleum. Although there are very extensive development efforts now underway in many countries to provide a basis for a new and improved commercial coal liquefaction industry, the basic technology involved is still based on the original German developments. A number of processes are under development in the United States such as the Gulf SRC-I and SRC-II, the Exxon EDS, COED/COGAS and H-coal. All of these processes produce a synthetic crude oil which can be refined into products somewhat similar to those that are produced from crude oil today, except that, in general, they will contain more aromatics. Conoco is involved in some of these projects and our subsidiary, Consolidation Coal, has been involved in coal liquefaction projects and investigations for more than 25 years. We have concluded from these many years of research that while these processes may eventually produce alternative fuels for coal, there is much work to be done in many unusual paths before they become commercial successes.

GASIFICATION

Liquids can also be made indirectly today using existing technology in which coal is gasified to make a mixture of hydrogen and carbon monoxide, and from that mixture, after suitable adjustments to composition, methanol, methane, or more complex mixtures of hydrocarbons can be synthesized using well-established and proven processes.

There are three basic types of gasifiers in which coal is contacted with steam and oxygen provides the energy needed for the gasification reactions. The three types are the concurrent entrained phase gasifiers, the countercurrent fixed or moving bed gasifiers, and the fluidized bed gasifiers. While all three systems may be operated at elevated pressure of 30 or more atmospheres, only the countercurrent fixed bed gasifiers, according to the Lurgi design, are presently proven and operate commercially at such high pressures. The commercial example of the concurrent entrained phase gasifier, the Koppers-Totzek, and the fluidized bed represented by the Winkler both operate at atmospheric pressure. Because of the power required to compress the gas leaving the gasifier to synthesis reactor pressures, little consideration is being given to the use of either of these gasifiers in indirect liquefaction plants.

THE LURGI GASIFIER

The most important gasifier available commercially today is that designed by the Lurgi Company in Frankfurt, Germany. It was initially commercialized in 1936 and since then many have been constructed throughout the world. Although some have been closed down due to age or because of the availability of cheap crude oil, the design is still being proposed for many synthetic coal-based alternative fuel plants in the United States, and it is the gasifier used in the very large complexes of SASOL in South Africa. Lurgi gasifiers are high pressure, fixed-bed, non-slagging, steam oxygen gasifiers and represent a wealth of commercial experience. This gasifier is, we believe, the only one which can be considered commercial today for the production of synthesis gas although there are a number of new systems which are proceeding at a rapid rate towards commercialization and may become available for second generation plants in the late 1980s.

The first of these is a modification to the Lurgi based on British Gas Corporation technology. This development began in 1955 but was stopped with the advent of cheap foreign oil imports into Britain. In 1974, as a result of the support of a group of companies based in the United States, work was begun in which a Lurgi gasifier was converted to the slagging mode. Since then, this gasifier has run through many tests, some supported by that group and later by the Department Of Energy, EPRI, the American Gas Association, and others. Indeed, one might say that at a level of consumption of coal of 350 tons per day (the size of the experimental gasifier), it could be accepted as commercially proven. In the next stage of the development of this gasifier, British Gas is building a 750 ton-per-day, eight-foot diameter version which is considered to be the largest that can operate using a single coal feed into a gasifier.

Texaco and Shell are both developing pressurized entrained phase coal gasification units. The Texaco process is a development for coal of a commercial process for gasification of petroleum residuals. The Shell process is a development designed to take full advantage of the commercial technology of Shell in pressurized oil gasification and the Krupp-Koppers atmospheric pressure coal gasification. Pilot plants, gasifying 150 tons-per-day, are currently in operation for both processes. A Texaco-based pilot plant has been in operation by Ruhrkohle at the Ruhr Chemie Chemical Complex in Oberhausen-Holten, West Germany, since January, 1978, and the Shell plant has been in operation since November, 1978, at the Hamburg refinery of Deutsche Shell.

There are a number of other gasification processes being developed in the United States and overseas. But these are at an early stage of development or are designed to operate under conditions which are not really suitable for the production of liquid fuel. Thus, if we look toward the production of alternative fuels from coal and we consider liquid as the main product, then the gasifiers that have been mentioned are the ones to be given prime consideration.

At Conoco we are convinced that the only gasifier that can be built and operate commercially today is the dry bottom Lurgi running on western coal, but it seems a reasonable probability that the BGC/Lurgi Slagging Gasifier, the Texaco, and the Shell gasifiers may be demonstrated at a sufficient size for the plants to be considered commercial later in the 1980s. Thus, we must expect that the first commercial production of coal liquids in the 1980s will be based on the Lurgi gasifier. By 1990, a number of gasifiers will be available which will provide an opportunity for eastern coal resources.

TYPES OF COAL LIQUID

The only large-scale commercial production of liquid from coal today is taking place in South Africa where the Fischer-Tropsch Process is used to synthesize large quantities of hydrocarbon fuels. This process is rather expensive, and except where very cheap coal is available is unlikely to be developed further. Much more likely to be developed is the production of methanol. Methanol is produced by synthesis from carbon monoxide and hydrogen, using proven processes, the best known of which are those developed by ICI and

Lurgi. Importantly, they all work effectively irrespective of the source of the synthesis gas, be it natural gas, naptha, heavy oil, or coal.

And why methanol? Methanol can be considered an environmentally benign fuel since it contains no sulfur and nitrogen and, therefore, burns without the emissions of any of the first and little of the second. Because it is a liquid, it can be burned in any piece of equipment which presently burns liquid fuel, and, therefore, we can consider many options for its use. It can be transported through a conventional pipeline distribution system, or by railroad and unit trains, and by water if a waterway is available. Thus, methanol can be made almost anywhere in the world and transported to a customer almost anywhere else in the world.

Methanol can be used in stationary engines, particularly for power generation; in agriculture as an alternative to liquid petroleum gas; in industry; and, most effectively, in automobiles. Methanol is a high octane fuel and, therefore, it is possible to construct an automobile with an engine somewhat different to one that uses gasoline. It can have a much higher compression ratio and run on a much leaner mixture, and as a result use 25 to 30% less fuel per mile.

It is difficult to conceive that any methanol will be produced via coal gasification before 1987, and we must expect the buildup over the first few years after this to be very slow. Investors are very wary of putting too much money into a new industry until they are satisfied that the first plants have operated satisfactorily. By the beginning of the 1990s, however, enough operating experience should have been gained to show that this is a satisfactory investment form. We can anticipate that between 1990 and the year 2000 a large number of plants will be constructed. The number will likely be constrained by the availability of A and E contractors, construction workers, and maybe even plant operators. But within these limits, we can anticipate that by the year 2000 there may be between 800,000 and 1,000,000 barrels a day of methanol being produced in the United States.

ILLINOIS COAL EXPORTS

So how can this affect the export of Illinois coal? Because it makes most sense to burn coal directly, encouragement must be given to develop those processes which enable high sulfur coal to be burned directly. Although stack gas scrubbing techniques are becoming available, users generally prefer to burn lower Btu low sulfur coal. Thus, the most important development to encourage the use of high sulfur coal is the fluid bed boiler which I mentioned earlier.

Thus, a simple analysis will tell us that there is a possibility of exporting Illinois coal indirectly by first converting it into methanol. The French government, with the announcement of its Carburol Programme last January, indicated the possibility exists. Other governments in Europe, notably West Germany and Sweden, have been supporting work on methanol as an alternative fuel for some years. Thus, the successful development of a methanol-from-coal industry in the United States could develop this indirect method of exporting coal. On the other hand, there may be those who object to this exploitation of U.S. resources in spite of the jobs it could create. Consideration must also be given to the limited capacity of the construction industry, as I mentioned earlier, which must raise serious doubts as to the availability of this type of resource for the benefits of other countries.

In the short term, perhaps we should invite them to cooperate with us to develop an alternative fuel industry, and for their share of the cooperation, allow them a share of the product. Then, after the successful commercialization, we should hope they will build their own plants and import our coal to use in them. During the time it takes to do this, we must improve our transportation system so that we can have a viable alternative to crude oil based on the world's largest energy resource--coal.

SYNFUELS TECHNOLOGY FOR HIGH SULFUR COAL
BERNARD S. LEE

Development of eastern energy resources could be easier than western resource development for a number of reasons. The eastern United States has the advantage of the following: 1) skilled manpower pools and manufacturing resources; 2) more water availability; and 3) much closer markets for synthetic fuels. The basic technical problem is to develop processes which will use the high sulfur bituminous coal generally found east of the Mississippi River. This problem can be solved by the development and use of processes which will produce synthetic fuels and recover sulfur as a by-product.

A number of options are available to produce clean gas and liquid fuels from coal (Figure 1). After preparation, the coal can be gasified using steam and air to produce a low Btu gas. This gas, after purification and sulfur removal, can provide a low Btu clean fuel gas with a heating value in the range of 100-250 Btu per cubic foot. A second option is to react the prepared coal with steam and oxygen to produce a medium Btu gas. This gas, after purification and sulfur removal, would be a medium Btu clean fuel with a heating value in the range of 250-550 Btu per cubic foot. This same gas could be methanated to produce a high-Btu gas (950-1000 Btu per cubic foot), which is completely interchangeable with natural gas. Another option is to use the gas as a feed for a Fischer-Tropsch synthesis to produce clean liquid fuels. The gas could also be converted into methanol, which is a desirable liquid fuel. The German-developed Lurgi Process (Figure 2) is commercially available and has been proposed for use in a number of coal conversion plants. However, the Lurgi Process is not suitable for gasifying Illinois coal.

HYGAS PROCESS

The HYGAS process (Figure 3) was developed at the Institute of Gas Technology (IGT) with funding from the American Gas Association and the federal government. The process was designed to use all types of coal feedstocks and to produce a pipeline-quality gas completely compatible with natural gas. The coal is pretreated and pumped as a slurry into a three-stage hydrogasifier. Steam and

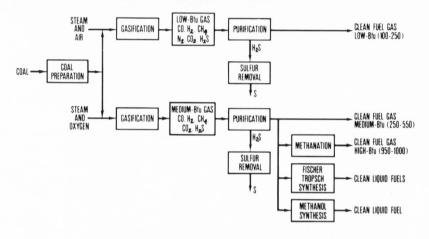

Figure 1. Clean Gas and Liquid Fuels from Coal

LURGI PROCESS

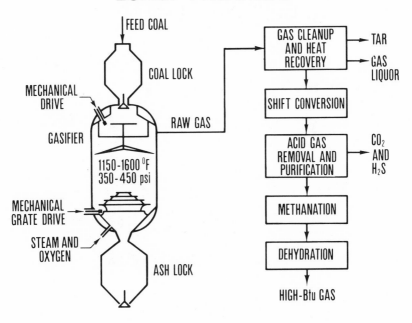

Figure 2. Lurgi Process

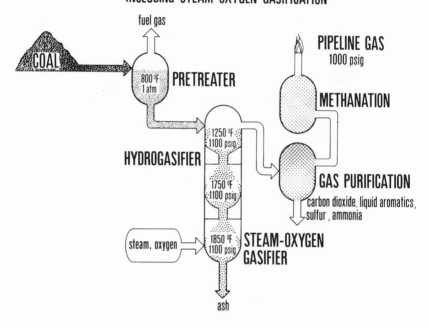

Figure 3. IGT Hygas Process

oxygen are introduced into the bottom stage and the resulting hot, hydrogen-rich gas reacts with the coal in the upper stages. The product gas, containing some methane, is purified by removal of carbon dioxide, liquid aromatics, sulfur and ammonia and is methanated to produce additional methane from the hydrogen and carbon monoxide in the purified gas stream. The product is pipeline gas which is completely substitutable with natural gas. The HYGAS coal-to-SNG process can use any U.S. coal, including Illinois Basin coal. A total of 22,700 tons of coal have been processed in plant operation exceeding 10,000 hours. About 267 billion Btu of gas have been produced, equivalent to 44,500 barrels of oil. Design for a commercial plant has been completed. The COGAS process is another advanced coal conversion system with high Btu gas and liquid co-products, and has been proposed for a plant in Perry County. At this time, plans to build demonstration plants using the HYGAS and COGAS processes have been suspended due to funding uncertainties.

U-GAS PROCESS

Medium Btu gas is another clean fuel which could be produced from Illinois coals. IGT has developed the U-GAS process to provide medium Btu gas for a number of uses. Figure 4 shows the configuration used to produce a medium Btu gas suitable for industrial fuel. One of the advantages of the U-GAS process is that the ash content of the coal is discharged as a dry granular material which would be environmentally benign. The characteristics of the process follow: 1) high conversion of coal to gas using the ash agglomerating technique; 2) capability to gasify all ranks of coal; 3) ability to accept fines in coal feed; 4) simple design and safe, reliable operation; 5) easy control and ability to withstand upsets; 6) production of product gas significantly free of tar and oils; and 7) no environmental problems. U-GAS pilot plant achievements include demonstration of process feasibility, utilization of different feed materials, sustained steady-state operations, the feeding of caking coals directly, and a coal utilization efficiency of 90 to 95%. The pilot plant has successfully operated on metallurgical coke, COED char, subbituminous coal, Illinois No. 6 coal and Western Kentucky No. 9 coal. The U-GAS process can also be used for combined-cycle power generation for electricity generation (Figure 5). Another potentially attractive configuration can produce a synthesis gas for use as a chemical feedstock (Figure 6). The Memphis Light, Gas and Water Division plans to construct an industrial fuel gas plant based on the U-GAS process and has applied to the U.S. Synthetic Fuels Corporation for a loan guarantee. The plant would use 3,100 tons

IGT U-GAS™ PROCESS
FUEL GAS GENERATION

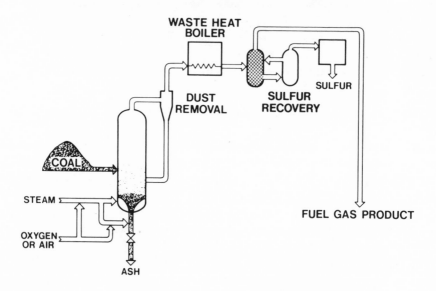

Figure 4.   IGT U-Gas Process
            Fuel Gas Generation

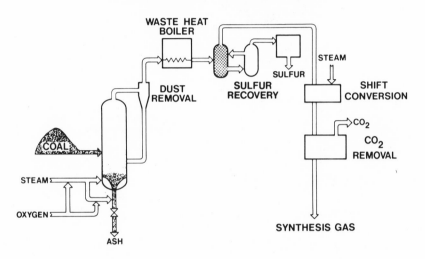

IGT U-GAS™ PROCESS
CHEMICAL FEEDSTOCK

Figure 5.   IGT U-Gas Process
            Chemical Feedstock

IGT U-GAS™ PROCESS
COMBINED CYCLE POWER GENERATION

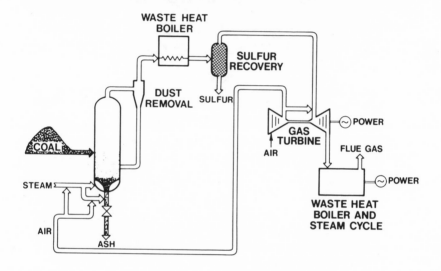

Figure 6. IGT U-Gas Process
Combined Cycle Power Generation

per day of western Kentucky coal and produce 50 billion Btu per day of 300 Btu per SCF of industrial fuel gas.

In conclusion, coal gasification offers a way to produce clean fuels from Illinois coals. A number of advanced processes are available and ready for demonstration. The processes are environmentally benign and their use on a large scale would be a major benefit to the state.

METALLURGICAL USES OF COAL
FROM THE ILLINOIS BASIN
MICHAEL HOLOWATY

The experience of Inland Steel Company with Illinois coals goes back to the early 40s. Pressed by the need for metallurgical coal for coke, Inland Steel resorted to the use of Illinois coals from Jefferson and Franklin Counties.

COKING BLENDS

Illinois coal as such is a marginal coking coal and its presence in the coal blends of that period of time resulted in reduction of coke quality. Over the years we developed our own method for coal preparation, tested it thoroughly, and increased the percentage of Illinois coal in coking blends to 35%. An Inland-developed flowsheet for coal cleaning was installed at our new mine in Sesser, Illinois, which went into operation in 1968. With this improved Illinois coal, we increased the percentage to 50%, the balance being West Virginia medium volatile coals. The introduction of the coal preheating technology in the 70s allows us now to use up to 75% of Illinois coal in our blend with the combination of two preheat and five conventional batteries. We are presently using close to 3 million tons of Illinois coal for metallurgical purposes. As we learned more about the properties of Illinois coal, we came to the conclusion that the optimum conditions would be attained if we could produce coke from 100% Illinois coal. The only way we could accomplish this was to produce formcoke.

FORMCOKE

This concept is entirely different from that of conventional by-product coke. Instead of slot ovens usually 18" wide, 20' high and 50' long, the formcoke plant resembles more a refinery with a number of cylindrical vessels and many transfer pipes. The coal is first devolatilized; then the resulting char is mixed with a binder produced from its own tar (pitch) and briquetted to any size and shape that might be required. The resulting briquette is then cured, which means hardened, and finally devolatilized in a vertical shaft furnace. The finished coke is ready to be used in the blast furnace in place of conventional by-product coke. At the present time there is one formcoke plant in Kemmerer, Wyoming, which has been in operation for 16 years. It is owned by the Phosphorus Division of FMC Corp. and produces formcoke and phosphorus reduction.

For our blast furnace test in 1973, which was carried out jointly with U.S. Steel, Armco, J&L McClouth, FMC and Inland Steel, we modified the Kemmerer plant to produce high quality 2" sized metallurgical coke briquettes and used 20,000 tons of this material in the No. 5 Blast Furnace of the Indiana Harbor Works. The test was a resounding success and the interest in the formcoke increased correspondingly.

In summary, the basic advantages of the formcoke process are as follows:

1. it can use single coal and almost any Illinois Basin coal can be used for formcoking;
2. it can be located at the mine-mouth;
3. it is environmentally clean; and
4. it can be automated to a great degree.

COKE SUPPLY IN THE U.S.

But now let us take a look at the coke supply situation in the United States through the eyes of Father Hogan of Fordham University, who recently completed a study for the Department of Commerce entitled "Analysis of the U.S. Metallurgical Coke Industry." In this study, Father Hogan writes the following:

During 1978, for the first time in nearly 40 years, the coke industry produced fewer than 50 million tons of coke, its output of 48.6 million tons falling 14.1% below the nation's consumption of 56.6 million tons. To offset the resulting shortages, consumers turned to overseas source of supply, and coke imports jumped from 1.8 million to a record 5.7 million tons or 10% of U.S. requirements....

The United States was able to import 5.7 million tons of coke because the industrialized nations of Western Europe, as well as Japan, did not experience rates of steel production comparable to those in this country; as a consequence, they had excess coke capacity available.

Dependence on foreign coke does not present undue supply problems when the steel industries in exporting countries are at relatively low rates of operation. However, when their operating rates increase to between 85% and 90%, their coke will be in tight supply, and, if it can be obtained from them at all, the price will increase significantly. During the 1973-74 worldwide steel boom, companies in the United States imported a substantial amount of coke for $125 per ton, which was over $50 per ton more than it cost to produce coke at that time in this country....

A detailed examination of the situation in these countries indicates that a revival of steel-mill operations would make it impossible for many to export coke and difficult for others. Thus, during periods of active worldwide steel activity, when U.S. coke production is inadequate to meet requirements, it would most likely be impossible to fill the void by imports.

Father Hogan then points out that 35% of the existing U.S. coking capacity of 22.4 million tons is over 25 years old. Another 20% of the capacity is 20-25 years old. As we assume the battery life at 30 years, then we can see that our coking capacity is aging very fast.

It would be well for the U.S. steel industry, with whatever help is necessary from the proper government agency, to examine the possibility of producing formed coke. This is a new process, still in the pilot stage, which permits the use of inferior coals and whose facilities are completely enclosed so that pollution would be reduced to a minimum. The claims for formed coke are indeed interesting and are to be seen. However, it is incumbent upon the industry and the proper government agencies to explore the possibilities of building a production-size facility in order to test the claims. If they prove to be valid, a substantial advance can be made in the use of coal reserves and pollution control.

What the Father Hogan study does not report is that much of the coke which we imported from Europe in the last five years and are importing even now was made with American coals transported across the ocean and back again because we have no coking capacity.

EUROPEAN COAL SUPPLIES

In the meantime, however, Europe is running out of coal. There are no operating mines in Sweden, Holland, Belgium, Italy, Spain, or Portugal. The French coal industry is expected to shut down within the next three years. The German and the British have sufficient coal reserves underground, but they are getting to be prohibitively costly to extract.

The question arises: Wouldn't it be a good prospect for us to produce coke in the United States and besides satisfying our own needs, export the coke instead of raw coal to Europe and other countries? If so, Illinois Basin coal converted into formcoke would be ideally suited for such a venture because of its access to the extremely economic river transportation system. The loading and unloading of coke briquettes presents no problem as was established in a 1972 study of the British Steel Corporation with the same kind of formcoke briquettes which we have used in our tests.

The formcoking process has one additional potential trump. The high sulfur Illinois Basin coal could be desulfurized in the char cooling stage, and thus it is conceivable that a 3% sulfur coal could be converted to coke with less than 1% sulfur. However, this will require additional work. How big is the market for such a coke product? I would estimate that it is upward of 10 to 20 million tons/year, of which about one half would go overseas and the other half would be used by our own steel industry. It should be borne in mind that the production of one ton of coke requires almost 1.7 tons of Illinois Basin coal.

Thus, in summary, what would be gained by establishing a strong coking industry in the Illinois Basin is this:
1. we would export a finished product instead of a raw material such as coal;
2. we would create construction jobs;
3. we would create operating jobs in formcoke plants, and additional jobs would be created in the coal mines and in the transportation;
4. we would build more efficient ports for exporting the cargos of coal and coke;
5. we would create, around the formcoking plants, industries supplying the formcoking plants themselves and other enterprises which would use the low Btu by-product gas from the formcoking operations in the production of glass, paper, textiles, etc.; and
6. we would beneficially affect our national balance of payments.

APPENDICES

APPENDIX A
ILLINOIS BASIN COAL AND THE WORLD EXPORT COAL MARKET

United States in the World Export Coal Market

    The United States has been the largest exporter of coal for many years and accounts for almost one-fourth of the world's export coal trade. The major portion of United States coal exports has been to the steel industry, which uses a high grade coal known as metallurgical or coking coal. The demand for this type coal is directly related to the demand for steel.

    A clear indication of the important role played by the United States in the total world market for export coal is presented in the following table.

Table A
ACTUAL COAL EXPORTS OF MAJOR EXPORTING COUNTRIES
(MILLION TONS)

Country	Year			
	1976	1977	1978	1979
United States	59.9	54.2	41.1	65.0
Australia	34.2	42.7	43.0	42.0
South Africa	6.6	14.0	17.0	25.0
Poland	43.1	43.1	44.7	45.0
Soviet Union	29.6	28.4	29.0	28.0

Source: Subhash Bhagwat, International Coal Trade: Trends and Practices. (Report to the Coal Advisory Committee to the Illinois Geological Society, 1980). Used with permission.

It should be noted that 1978 United States export totals were dramatically reduced by the 110-day United Mine Workers' strike.

    According to WOCOL, the United States will be capable of satisfying over 35% of the future export coal demand. This amounts to a potential of 221 to 386 million tons of United States export coal. The United States, in addition to being a major metallurgical coal exporter, is also expected to play a dominant role in satisfying the increasing worldwide demand for steam coal. Export steam coal demand in the electric utility market will become more prevalent because of constraints on the availability of oil.

Two of the most prominent factors that will determine the level to which the United States participates in the growing world market are the conditions and capabilities of present and future port facilities. An adequate port infrastructure is a necessary link in meeting the demand to supply the world with United States coal. Present transshipment facilities, located almost exclusively on the East Coast, are barely adequate to meet today's needs, and are being strained by the rapid increase in coal tonnage moving through the facilities. Tremendous congestion and lengthy delays are occurring as a result of the lack of ground storage, blending activities, and the sheer volume of tonnage being handled. These delays, at the East Coast ports of Baltimore, Norfolk, and Newport News, are causing ships to incur demurrage charges of $15,000 or more on the delivered price of coal. Many of the older ports have historically been geared more toward loading low-volatile metallurgical coal, which requires special handling, rather than for the high volumes of steam coal being handled today.

A recent New York Times article on the East Coast port problems accurately depicted the major problems confronting coal exports through eastern ports. A common sight is many vessels offshore awaiting berth availability. For example, on September 11, 1980, 51 ships were anchored in the Hampton Roads area awaiting coal loading. Many of these vessels waited from 3 to 4 weeks before being loaded. Total 1979 coal exports through various port facilities are shown in the Table B.

Table B
UNITED STATES OVERSEAS EXPORTS OF BITUMINOUS COAL
BY PORT OF EXIT:
TOTAL EXPORTS EXLUDING GREAT LAKES PORTS

Port of Exit	(000's tons)	% of Total
Baltimore	10,184	17.2%
Hampton Roads	45,753	77.5
Mobile	1,641	2.8
New Orleans	1,410	2.4
Philadelphia	55	0.1%
Total	59,043	100.0%

Source: Mark E. Tomassoni, "Coal: The Pressure's on Ports," American Seaport (August, 1980), p. 8. Used with permission.

The two major port facilities at Hampton Roads, Virginia, handle 77% of total United States coal exports. The Chessie system owns and operates Piers 14 and 15 at its Newport News, Virginia, Terminal. Coal is stored in railroad equipment in their yard, which can hold up to 6,500 loaded cars. Additional track storage is also available nearby. Water depth at the Pier 14 berth is 45 feet and 38 feet at Pier 15. Coal is loaded into vessels at a rate of up to 9,000 tons per hour. The Chessie estimates that 15 million tons of coal will be handled through Newport News in 1980. The maximum capacity is about 20 million tons per year.

A. T. Massey, a large eastern coal broker, has purchased the Chessie system's Pier 9 at Newport News and will construct a modern coal terminal at this site. When completed, the terminal will have a throughput capability in excess of 10 million tons annually and will alleviate some of the congestion now being experienced at Hampton Roads. Total construction cost is estimated at $53 million.

The other terminal at Hampton Roads is Lamberts Point, which is owned and operated by the Norfolk & Western. Pier 6 is the world's largest coal-handling facility. The maximum operating rate is 16,000 tons per hour when two vessels are being loaded simultaneously. Storage is in railroad cars and the rail yard has a capacity of 15,600 cars. Water depth at Pier 6 is 46.5 feet. The Norfolk & Western originates coal movements from over 200 mines in West Virginia, Virginia, Ohio and Kentucky. In 1980, over 30 million tons of coal will be handled through Lamberts Point, and less than 10% will be steam coal.

Conrail's throughput is nearly 2 million tons per year through the Port of Philadelphia Pier 124. Conrail is developing plans to modernize existing facilities and expand to a new terminal in the Delaware Port Terminal area. Conrail's modernization program will increase coal-handling capacity to 10 million tons annually and will be completed in 1982[1].

The Chessie System also operates a pier at Curtis Bay in Baltimore, Maryland. In 1979, 17% of United States coal exports were shipped through Baltimore. The Chessie can store 4,700 cars of coal at the terminal. Water depth is 42 feet on the south side of the pier, and 30 feet on the north side. Total export capacity is estimated at 12 million tons per year. A temporary shutdown last year for maintenance contributed to the already growing problem.

The New Orleans Public Bulk Terminal, with existing open ground storage for 85,000 tons of coal and expansion capacity for up to 150,000 tons, is located on the Mississippi River Gulf outlet (Industrial Canal) and currently provides the Port of New Orleans with coal-exporting facilities. Current channel depth is 36 feet, which can accommodate vessels up to 40,000 deadweight tons (DWT).

The Bulk Terminal can load vessels at a rate of 2,000 tons per hour and has a maximum annual throughput of about 4 million tons.

The Port of Mobile, Alabama, has a bulk terminal operated by McDuffie Terminals, Inc., which is now capable of handling approximately 3 million tons of coal per year. When current expansion plans are completed, McDuffie Island will be able to handle up to 7 million tons of coal per year. The river channel is 40 feet deep and open ground storage will accommodate 500,000 tons of coal. Vessels are loaded at a rate of 4,000 tons per hour.

Long Beach, California, is handling a few test shipments of export coal originating in the western coalfields of Utah and Colorado. Open storage today is 50,000 to 80,000 tons. Vessels are loaded at a rate of 1,200 tons per hour. The channel depth is 40 feet.

Additional investments will be necessary to modernize and enlarge the actual transfer facility equipment in the United States to be able to compete effectively in the world marketplace. Deepening of many ports and channels will enable newer and larger coal vessels to be used, thus creating greater economies of scale. Significant ground storage will enhance a port's ability to attract the available export coal business. Increased ground storage will also reduce coal storage in railroad cars, thereby minimizing car requirements and investment.

ILLINOIS BASIN IN THE WORLD EXPORT MARKET

Availability/Supply

The Illinois Basin, with its massive coal reserves, is capable of playing a vital role in meeting worldwide coal demands. The Illinois Basin coalfields are located in Illinois, Indiana and western Kentucky. Counties served by the Illinois Central Gulf contain approximately 115 billion tons of recoverable coal reserves.

Historically, the United States has exported metallurgical coal to the world's steel-making industry. Because most metallurgical coal is located in the East, Illinois Basin producers have not been a significant factor in the export coal market.

Reduced demand for metallurgical grade coal has led to oversupply situations, resulting in reduced prices; this, in turn, has allowed metallurgical coal to move into steam coal markets. Metallurgical coal is desirable as a utility fuel because of its high Btu level and low ash and moisture content. These conditions have made it difficult for Illinois Basin producers to penetrate the export steam coal market.

When metallurgical coal demand increases with future growth in the steel-making industry, prices will increase, thereby making metallurgical coal uneconomical for use as steam coal. Demand by utilities for a heating fuel will then shift back to steam coal, including that which is available in the Illinois Basin.

Supply/Reserves

Within the Illinois Basin, there are more than 215 billion tons of identified coal reserves[2]. Of these total coal reserves, better than 75% are known to be within the State of Illinois--161 billion tons. Of this amount, 115 billion tons of recoverable coal reserves are located in counties served by the Illinois Central Gulf. Illinois contains more coal reserves than any other state, except Montana.

Coal reserves in the Illinois Basin are found in large continuous blocks. This concentration of coal reveals 19 Illinois counties with over 3 billion tons each, and 13 of these counties are served by the Illinois Central Gulf Railroad.

Because of this concentration of coal reserves, any expanded coal production in the Illinois Basin is expected to come from Illinois coal mines.

Capacity of Producers

In retrospect, it appears that the 1973 oil embargo, far from being a panacea for the coal industry, was instead a damaging event from which the industry has yet to recover. Margins are still low as a result of depressed prices, declining productivity, and excess capacity. Most authorities estimate that current excess coal capacity is more than 100 million tons annually[3]. The estimated oversupply breakout by region is shown in Table C.

Table C
SURPLUS COAL BY REGION

Territory	Tons (Millions)
Eastern Coal	35
Illinois Basin Coal	15
Western Coal	50
TOTAL	100

Production capacity of the coal industry continued to climb as producers, convinced that the coal boom was soon to come, increased production. Production was static at approximately 600 million tons annually from 1970 to 1974, then grew to nearly 700 million tons in 1977. By 1979, it had grown to 775 million tons and is expected to reach 815 million tons in 1980. Industry analysts predict that the coal industry will continue to experience excess capacity and depressed prices through the early 1980s.

The impact of these trends has been felt nationwide, especially in the Illinois Basin where coal mines have closed as a result of low profitability, excess production, and restrictive Environmental Protection Agency standards. This has reduced marketability of this type of coal. Many mines that were opened in the 1960s produce high sulfur coal, which was adequate for commercial use at the time, but now has limited marketability because of Environmental Protection Agency emissions restrictions. Slackened industrial coal demand also has diminished the potential to sell coal in the open market. Other mines are closing because of restrictions imposed by government regulation on strip mining reclamation.

Coal produced in Illinois and western Kentucky has declined in recent years as a result of these factors. For example, in 1974, Illinois mines produced 58 million tons of coal, but only produced 49 million in 1978. Western Kentucky mines produced 52 million tons of coal in 1974, but only 39 million tons in 1978.

More than 30 coal producers operate in the Illinois Basin. Much of this coal production comes from mines operated by subsidiaries of large United States corporations, which provide a strong financial base vital to future expansion plans. Examples of this ownership are depicted in Table D.

Table D
OWNERSHIP OF ILLINOIS BASIN COAL MINING COMPANIES

Mining Company	Parent Company
AMAX Coal Company	AMAX, Inc.
Old Ben Coal Company	Sohio Petroleum Company
Consolidation Coal Company	Continental Oil Company
Freeman United Coal Mining Company	General Dynamics
Island Creek Coal Company	Occidental Petroleum Corporation
Zeigler Coal Company	Houston Natural Gas Corporation
Arch Mineral Corporation	Ashland Oil Company/ Hunt Petroleum Corporation
Pittsburgh & Midway Coal Mining Co.	Gulf Oil Company
Sahara Coal Company, Inc.	Sahara Coal Company
Inland Steel Coal Company	Inland Steel Company
Kerr-McGee Coal Corporation	Kerr-McGee Corporation
ARCO Coal Company	Atlantic Richfield Company

Source: George F. Nielsen et al., <u>1980 Keystone Coal Industry</u>
<u>Manual</u> (New York, NY: McGraw-Hill Mining Publications,
1980), p. 710. Used with permission.

Future production from Illinois Basin mines will have to come
from deep mines which are capital-intensive and technically far more
complex than surface mining. However, the operating experience and
financial strength these possess will provide a distinct advantage
to the future development of Illinois Basin coals.

Quality
    A primary concern of utilities in purchasing coal is its sulfur
content. The sulfur content of United State coal reserves is
indicated in Table E. Western coal reserves are primarily low
sulfur (less than 1%) whereas the sulfur content of Appalachian coal
reserves ranges from low to high. The bulk of midwestern reserves
are high sulfur (greater than 3%).

Table E
UNITED STATES COAL RESERVES--SULFUR CONTENT

100	12%[a]	23%[a]	6%[a]
			High 1%
75	High 21%		Medium
		High 61%	
50	Medium 39%		
			Low 77%
25	Low 28%	Medium 14%	
0		Low 2%	
	Appalachia	Midwest	West

Note: Low sulfur = 1% or less
       Medium sulfur = 1.1 to 3%
       High sulfur = 3% or more
       [a] = Unknown

Source: From <u>Future Coal Prospects: Country and Regional</u>
          <u>Assessments</u>, Copyright 1981, Massachusetts Institute of
          Technology. Reprinted with permission from Ballinger
          Publishing Company.

Table F summarizes the quality characteristics of representative United States coal reserves by state, coal type, Btu
content, and sulfur and ash percentages.

Table F
CHARACTERISTICS OF REPRESENTATIVE COALS

State	Coal Type	Btu/lb.	% Sulfur	% Ash
Pennsylvania	Bituminous	12,067	2.03	15.0
West Virginia	Bituminous	12,516	2.39	12.1
Ohio	Bituminous	11,047	3.42	15.8
Kentucky (East)	Bituminous	11,784	1.23	12.5
Kentucky (West)	Bituminous	11,990	3.15	9.5
Illinois	Bituminous	10,775	2.92	11.4
Alabama	Bituminous	11,740	1.43	14.1
Colorado	Bituminous	10,925	0.49	9.4
Wyoming	Subbituminous	9,037	0.50	8.3
Utah	Bituminous	11,569	0.63	11.9
Montana	Subbituminous	8,957	0.64	7.6

Source: Federal Energy Regulatory Commission. <u>Annual Summary of</u>
          <u>Cost and Quality of Electric Utility Plant Fuels, 1977.</u>
          1978.

Major concerns regarding Illinois Basin coal center around its
quality and sulfur content. General characteristics and qualities
of Illinois Basin coal include a Btu range of 10,100 to 12,700; a
sulfur content range from 1.0 to 5.0%; an ash content of 5 to 7%.
This coal is basically of high quality for steam coal purposes,
except for the generally high sulfur content.

Approximately 5 billion tons of total Illinois coal reserves
have a sulfur content of 2% or less[4]. These reserves are of
acceptable grade for steam coal and will play an important role in
future Illinois Basin export when mines are developed.

Competitive Environment

Geographic Factors. Illinois Basin refers to coal seams in
Illinois, Indiana, and western Kentucky. The major rivers in the

area are the Mississippi, Illinois, Ohio, Tennessee, Green, and Kaskaskia. Surrounding the Basin are several major cities, including Chicago, St. Louis, Louisville, and Indianapolis.

Competitive Pricing. Illinois Basin coal can be competitive in the world export market. Each link in the production and transportation chain is competitive, including the minehead price of coal, rail rates, handling charges and vessel rates. The Illinois Central Gulf's role in this chain is as a transportation company offering competitive freight rates and service. The consignee is assured that the Illinois Central Gulf's price will remain competitive as a result of the automatic pressures from East Coast producers and by inland water carriers on the Mississippi River system. In fact, Illinois Basin coal for export may have the edge in certain areas. First of all, total transit time to New Orleans from the Illinois Basin will be faster via the Illinois Central Gulf than via barge. Second, the Illinois Central Gulf main line is in good physical condition and is ready to handle direct movement of coal unit trains from the Midwest to the Gulf. Additionally, economical and demand-sensitive unit train rates can now be offered. The Interstate Commerce Commission has recently granted the Illinois Central Gulf permission to compete readily for the export market by authorizing it to raise or lower rates on two days' notice to meet competitive conditions[5].

Environmental Factors. The higher sulfur content of Illinois Basin coal, from many of the existing mines, often restricts its use today in various markets. Together with the United States, Europe and Japan have strict environmental standards limiting the burning of high sulfur coal. The sulfur maximum in these countries is 2% and preferably less. Blending of Illinois Basin coals at the port of exit or at destination is considered a realistic means to meet stringent sulfur laws and should be tested to prove this hypothesis.

World Market Competition. The United States, China and the Soviet Union hold the majority of world coal reserves (62.5%), with Australia, South Africa, Canada and Germany holding most of the balance. France, Italy and Japan are the major importers. The export steam coal market is expected to rise dramatically as a percent of total world exports, placing the Illinois Basin, which has large reserves of steam coal, in a strong position to compete effectively. Because of its vast resource base, the financial strength of the coal producers, and an excellent transportation infrastructure, Illinois Basin coal can become competitive with other exporting countries and with eastern and western United States coal entering the export market.

FACTORS AFFECTING EXPORT OF ILLINOIS BASIN COAL

Coal Producers

WOCOL has documented the potential of the steam coal export market. Some producers in the Illinois Basin are beginning to utilize long-range strategic planning to meet the anticipated export demand. This strategic planning includes a major effort to identify low sulfur coal deposits and create an effective marketing campaign to describe the advantages of using Illinois Basin coal. Reactions from Illinois Basin coal company executives are positive and many are actively engaged in marketing their coal for short and long-term buyers.[6] This is a major change of policy from the reactive marketing posture of the recent past, demonstrated by many companies in connection with the domestic utility market.

Regulatory Environment

The short-term demand (1980-1983) for Illinois Basin coal in the export market may be limited by several environmental factors. Japan and the European importing countries, major market targets for United States steam coal, have strict environmental regulations limiting coal utilization to those grades low in sulfur content (2% or less). Presently, several existing mines in the Illinois Basin region are capable of meeting this coal grade requirement. These mines include Freeman United (Orient III and VI Mines), Old Ben Mine No. 21, Inland Mine and Sahara Mine, all located in southern Illinois, and located on the Illinois Central Gulf. Generally, west Kentucky coals are not in this required sulfur range. However, future blending of various types of coals may alleviate this problem.

Limiting Factors

The vast majority of potential importers believe that Illinois coal has a high sulfur content, ranging from 3 to 5%. Illinois Basin coal producers must educate buyers and potential buyers that there are many grades of Illinois Basin coal available, including considerable low sulfur coal.

New Technology

Several new coal-using technologies are also expected to enhance the demand for Illinois Basin coal. These technologies include fluidized bed combustion, coal-oil mixtures, coal gasification and coal liquefaction.

One of the more promising new technologies, fluidized bed combustion, offers a number of important advantages over conventional boiler utilization. This process absorbs sulfur oxides and minimizes the problem of removing them from the flue gas; allows the efficient boilers. However, expanded research and development of these technologies is required, which will delay commercial application perhaps 10 to 20 years.

Transportation Requirements and Capabilities

In terms of transporting Illinois Basin coal to the export market, the rail system and the water system are in place. The Illinois Central Gulf main line is in good condition to handle unit trains from Illinois to the Gulf and, more importantly, is a direct north-south route.

Transit time to the Gulf from the Illinois Basin via railroad unit trains is approximately 2-3 days, while barges can take as long as 2 weeks. The elimination of the rail-to-barge or truck-to-barge transfer stage will also save time and expense when shipping all-rail.

As a rail transportation company, very little additional investment by Illinois Central Gulf is necessary to handle a large increase in export coal traffic.

ICG'S VIEW OF FUTURE EXPORTS

With proper planning, the necessary coal mines, transportation facilities, and construction of modern colliers can be readied to handle future coal exports. Such planning will necessitate close cooperation among coal producers, transportation companies, and the coal-importing countries.

The long-term demand for Illinois Basin coal in the export market is more optimistic than that found in the short term. Technical advances in the control of particulate and gaseous emissions will be required to meet increasingly strict environmental regulations. Generally, the importing countries do not have any less restrictive environmental regulations than the United States. In this area, considerable progress has been made in recent years to provide the purchasers with the base mix of equipment, price, availability and reliability of pollution control equipment. Progress in the area of emission controls and sulfur removal is of particular importance to the marketability of Illinois Basin coal because of its level of sulfur.

Marketability of Illinois Basin coal will be enhanced by new technology, including advanced scrubber systems, improved washing techniques, blending of various grades of coal, and other technological developments.

Coal ownership by major oil companies has brought vast capital resources and management expertise to the coal industry in the Illinois Basin, which is vital to its expansion.

The Illinois Basin export steam coal market holds the potential for high return at relatively low financial exposure to the Illinois Central Gulf. The current and potential production capacity of the Illinois Basin producers, coupled with the 115 billion tons of coal reserves, including the 5 billion tons of low sulfur coal reserves, will provide the foundation for Illinois' participation in the export coal market.

NOTES

[1] "Conrail to Expand Export Facility." Coal Mining and Processing, 17, No. 19 (September 1980):31.

[2] Nielsen, George F. et al. 1980 Keystone Coal Industry Manual (New York: McGraw-Hill Mining Publications, 1980), p. 497.

[3] The Coal Industry Quarterly Review (New York: Merrill Lynch, Pierce, Fenner, and Smith, Inc., August 1980).

[4] Simon, Jack A. and Hopkins, M. E. Coal Resources of Illinois (Illinois State Geological Survey, Illinois Minerals, 1974).

[5] Illinois Central Gulf Railroad Coal Tariff 4163 (Chicago: Illinois Gulf Central Railroad, October 7, 1980).

[6] Interviews with executives from Freeman United Coal, Kerr-McGee, Old Ben Coal, Island Creek Coal, and U.S. Steel.

APPENDIX B

THE GROWING INTERNATIONAL STEAM COAL TRADE:
AN OUTLINE OF SEVEN STRATEGIC CONSIDERATIONS
AND RECOMMENDATIONS FOR MIDWESTERN COAL PRODUCERS

I. Clarify and understand the implications of coal-exporting objectives.

  A. Generally, U.S. firms have never been as export-oriented as their counterparts in Europe or Japan. It is easy to understand the reasons for this situation: a limited metcoal capability, poor port facilities, and uncertain logistical, market, and financial factors.

  B. But today there is a tremendous and increasing opportunity for steam coal exports. However, to take full advantage of this development requires possible significant changes in the structure and operations of midwestern coal producers. Therefore, the producers must determine how much they are willing to change in order to capture a portion of the coal export market.

II. Define the dimensions of export production and midwestern distribution infrastructure. More specifically, the coal producers must assess their own production capabilities. The firm should establish:

  A. whether it can produce coal for export (now, in the future, and at what levels); and

  B. whether it can produce a regular flow of coal for a number of years (can the producers, for example, establish a certain percentage of their production for export?).

III. Assess the coal export markets.

  A. Gather information on coal-importing nations.

  B. Determine overall energy demand (especially for electricity, e.g., the likely prospects for nuclear).

  C. Assess overall coal demand.

  D. Identify the critical sectors for coal import with respect to:

    1. short term (e.g., the cement industry in Spain); and

    2. long term (e.g., electric utilities generally).

  E. Assess the importing nation's infrastructure in detail, especially port facilities and ground transportation

(railroad and truck). Remember that at the operating
level, coal-importing countries can have drastically
different procedures from the U.S.: port operations can
be dominated by unions, or other groups, which, in turn,
might be closely connected to political parties or ide-
ologies, having reactions that cannot be analyzed on
purely economic terms; police and judicial systems might
not be cooperative in some cases; in other cases companies
operating in ports (unloading, transporting, transshipping, storing, etc.) might not be as reliable as they
should, because of bureaucracy and slower services (bank-
ing, customs, insurance, brokerage, etc.); and processes
might have delays. Such problems have been solved in many
cases by exporters, importers, and traders, even in
difficult locations such as Nigeria or Saudi Arabia, where
ports are usually jammed, or at Mediterranean ports where
labor disputes are normal. Often, solving these problems
does not require large sophisticated organizations. It is
enough to have one or two people, knowledgeable about the
details of these logistical processes, who can identify
and coordinate the adequate local services.
F. Establish the cost-benefit exchange for a nation's coal
   market: Is pursuing a coal market worth the time,
   resources, and risk?

IV. Assess the domestic infrastructure.
   A. Determine current port capacities and future capacities.
   B. Judge ground transportation: railroads, barges, etc.
   C. Establish overall, realistic current capacities, forecast
      future improvements, and target weak links in the system.

V. Address specific possible actions for participation in the coal
   export trade.
   A. Establish effective channels of distribution overseas:
      Figure 1 shows a variety of possible channels of
      distribution for coal. The coal producer essentially must
      determine whether he wants to be his own seller or to
      allow a third party, such as a trader, to sell the coal.
      There are costs and benefits for either approach. Also,
      it is important to note that although many importing
      countries have coal-buying "monopolies," the actual users
      also buy coal directly. Other critical issues related to
      channels of distribution include the following:
      identification of critical elements to control, methods of
      control (solely owned, joint venture, subcontract, etc.);

means of selling (long term or spot); and determination of appropriate level of risk (demurrage, devaluation, etc.).

B. Establish an effective ongoing intelligence and data-gathering system.

C. Perhaps it is even possible to establish market share criteria such as goals, i.e., capturing a targeted percentage of a nation's coal import needs (as is done in other industries).

D. Be prepared to establish a long-term presence in Spain (or Italy): avoid strictly short-term criteria or goals for success (remember the Europeans have already been burned).

E. Or alternatively look for viable, credible partners (e.g., scrap dealers) as at least one way to be considered trustworthy and legitimate (note: again this calls for accurate, detailed, country intelligence).

F. Or alternatively establish a small permanent office in Spain (Italy); (again, we are taking the long view).

G. Work with or through effective and trustworthy Spanish (Italian) intermediaries.

H. Effectively use midwestern trade groups, the railroads, and the port authorities (note: use all U.S. parties that would gain by exporting more coal); this kind of action could be especially useful in improving the U.S. infrastructure.

VI. Determine essential internal actions required (note: this is critical, for the U.S. company needs to get its own house in order).

A. Develop the organization.

    1. Top level involvement is essential.

    2. Perhaps one should establish a permanent coal-exporting project group (by country?).

    3. Educate key internal people throughout the firm.

B. Develop a working plan.

    1. Identify key customers.

    2. Identify methods for reaching customers (see Figure 1).

    3. Develop public information material about the region's coal and develop ways to establish credibility.

    4. Develop selling materials on producer's own coal (brochures, current data).

    5. Be knowledgeable on U.S. infrastructure.

    6. Develop coalitions and allies in the U.S.

    7. Begin carefully and do not promise unless delivery is certain.

Figure 1
ALTERNATIVE CHANNELS OF DISTRIBUTION FOR THE
INTERNATIONAL COAL TRADE

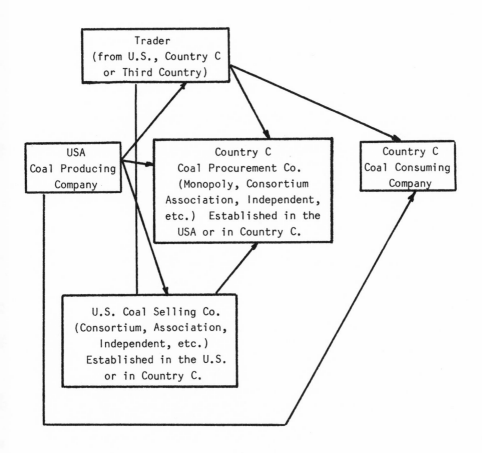

8. Be ready to adapt as new information emerges.

VII. The following represent overall elements of success:
   A. A long-term perspective in relation to the steam coal international coal trade is vital.
   B. It is necessary to have a permanent presence in some form overseas in selected markets.
   C. One must understand the coordination of the infrastructure in the U.S.
   D. Also important are managerial change and creativity within the firm.

APPENDIX C
RECOMMENDED FEATURES OF DEEP DRAFT LEGISLATION

A number of bills have been introduced (and reintroduced) recently in both the Senate and the House to authorize a fast-track procedure for deep draft dredging projects, both for named ports and, pursuant to criteria applicable, to all ports. The Port of New Orleans feels that any deep draft legislation finally enacted should contain the features in the following outline:

I. Fast-Track Procedure
    A. To avoid the delays inherent in the normal course of the environmental permit process, the legislation should provide for submission to Congress of an EIS addressing the purposes, policies, and requirements of the various environmental laws, with the further provision that, absent congressional disapproval by concurrent resolution within 60 days, the requirements for these environmental laws will be considered as satisfied.
    B. With respect to ports not named in the authorizing legislation itself, provision should be made for congressional approval of the navigation improvement (as satisfying the generic criteria required for use of the fast-track procedure) either 1) as part of congressional approval of the EIS for the project, or 2) by provision for a threshold submission to Congress, with further provision that, absent congressional disapproval by concurrent resolution within 60 days, the navigation improvement will be considered as authorized.
    C. The legislation should follow the provisions for judicial review in the Trans-Alaska Pipeline Act, 43 USCA S1652 (g), in 1) restricting judicial review to allow only the assertion of claims alleging the invalidity of the Act, the denial of constitutional rights under the Act, or action in excess of authority under the Act; in 2) requiring the filing of such claims within 60 days of the action in question; in 3) precluding the issuance of injunctive relief except in connection with a final judgement; and, in 4) allowing direct appeal to the Supreme Court.

D.   The legislation should <u>not</u> allow the assertion of claims alleging the inadequacy of the final EIS <u>after the EIS is considered approved by Congress</u>. Such a legislative finding and determination should not be subject to judicial review by claimants who disagree with the legislative wisdom of the decision. Congress' judgment should remain final, except as limited by the Constitution.

E.   The provisions for judicial review should be the sole and exclusive means for review of navigation improvements under the Act.

II. Provisions for Cost Recovery

A.   Recovery of federal costs for construction and O & M should be limited to those costs attributable to the channel deepening authorized by the legislation, and should not include costs associated with previously authorized channel depth and maintenance. To provide otherwise would go beyond requiring local assuring agencies to bear their share of the navigation improvement in question and would have the effect of imposing additional costs upon ports for previously authorized and undertaken improvements.

B.   Recovery of federal costs should be limited to 25-50% for construction and O & M.

C.   Recovery of federal costs from a local assuring body should be limited to "net federal costs," which should be defined as construction costs and O & M costs <u>less</u> benefits accruing to areas, interests, or persons beyond the local assuring body (e.g., regional and national benefits).

D.   The cost-reimbursement obligation of non-federal interests should be accompanied by a corresponding right to recoup reimbursement costs by the imposition of user charges and fees, and the reimbursement obligation should be binding upon the local assuring body (and payable) only to the extent of user fees and charges collected.

E.   Assurances of local cooperation required for the navigation improvement should be such assurances as are "satisfactory to the Secretary of the Army."

F.   Navigation improvements required as a result of extraordinary and unforeseen forces (e.g., <u>force majeure</u> or <u>vis major</u>) should be excluded from any cost reimbursement obligator.

APPENDIX D
PRIMARY BENEFITS OF A 55-FOOT CHANNEL FOR
THE MISSISSIPPI RIVER MOUTH OF PASSES
TO PORT OF BATON ROUGE

EXECUTIVE SUMMARY

The deepening of the channel of the lower Mississippi River between the Mouth of Passes and the Port of Baton Rouge will prove to
be of significant benefit to the petroleum and petrochemical
industries in southeastern Louisiana. Larger ships carrying greater
payloads will be able to navigate the river. This will actually
decrease the number of ships on the lower Mississippi, and therefore
it will improve the operational and safety characteristics of the
remaining traffic. The larger ships will transport cargo more cost
effectively. In fact, their use will result in significant savings
to local consumers of commodities which are transported by ship.
The petrochemical and petroleum industries could have realized a
savings of up to $228,300,000 in 1980 if imported crude and
petroleum products had been shipped to their ports of destination in
150K DWT tankers rather than in smaller ships and barges. This
figure, which is substantiated in the body of this report,
represents savings of approximately 43% of the total transportation
costs of crude from selected ports of origin. Given the critical
need for petroleum and petrochemical products in the United States,
the deepening of the navigable channel of the lower Mississippi to
55 feet will prove to be of national significance and will serve to
maintain the ports of New Orleans and Baton Rouge as major world
ports.

PRIMARY BENEFITS

Deep draft vessels (Very Large Cargo Carriers--VLCC) will be
able to navigate the lower Mississippi River. The salient benefits
of this will be several, as indicated by the following:
1. It will reduce river traffic through the use of larger
    ships with greater payloads. This reduction will facilitate
    the operating characteristics of traffic on the river and
    actually provide additional space for a net growth in cargo
    transported.

2.   Larger vessels, specifically tankers, operate in a more
     cost-effective manner than smaller ships. A review of current
     World Scale rates demonstrates that ships in he 150K DWT range
     offer a transportation savings of up to 43% over those in the
     80K DWT range. Given the amount of crude that will be
     transported into this area, the savings realized through the
     use of VLCC will be substantial. If all crude and No. 6 fuel
     oil imported in 1980 had been shipped in 150K DWT tankers
     rather than in 80K DWT tankers, $228,301,600 would have been
     saved.

3.   The need to lighter crude will be substantially reduced,
     if not eliminated. The cost of transferring oil from ship to
     barge has grown to approximately $1.00 a barrel of crude
     lightered, and lightering has often resulted in unanticipated
     delays and demurrage charges on the affected tankers.

4.   The elimination of the constraint of the 40-foot channel
     may encourage more crude to flow into the area; that is, local
     refiners will be able to search more widely in world markets
     for crude because of improved access to their ports of
     destination. Transportation costs will be reduced, and thus
     distant markets will prove to be more competitive. This
     benefit will be realized throughout the American economy
     because products are sold in domestic markets. The lowered
     cost of production will be reflected in the price of refined
     products.

5.   The 55-foot channel will prove to be a cost effective alterna-
     tive to the Louisiana Offshore Oil Pipeline for those users
     (refiners) who are not members of LOOP. The presence of an
     economical alternative mode of transportation will allow
     independent refineries to maintain a competitive posture with
     the major oil companies in LOOP. The presence of competition
     in the market will assist in stabilizing the cost of products
     to the American consumer and industrial users.

                         SUMMARY STATEMENT
              TRANSPORTATION OF PETROLEUM AND RELATED
              PRODUCTS ON THE LOWER MISSISSIPPI RIVER

     Petrochemical movements with points of origin or destination on
the lower Mississippi River between the Mouth of Passes and the Port
of Baton Rouge totalled 170,609,541 short tons in 1978. The
transportation of crude petroleum accounted for 106,355,853 short
tons or 62.3% of this tonnage (see Table 1, next page).

Table 1
1978 PETROCHEMICAL MOVEMENTS ON THE MISSISSIPPI RIVER
FROM MOUTH OF PASSES TO THE PORT OF BATON ROUGE
BY COMMODITY AND SHORT TON

		FOREIGN	
COMMODITY	TOTAL	IMPORTS	EXPORTS
Crude petroleum	106,355,853	78,378,705	1,093
Basic chemicals	8,401,042	187,886	1,831,526
Gasoline	15,624,887		25,533
Distillate fuel oil	10,279,335		132
Residual fuel oil	24,039,569	34,906	
Lubricating oil	3,074,192	75,510	633,839
Coke, petroleum coke	2,829,663	1,634,609	256,962
GRAND TOTAL:	170,609,541	80,311,616	2,747,085

SOURCE: Department of Army Corps of Engineers, "Part 2: Waterways
and Harbors-Gulf Coast Mississippi River System and
Antilles," Waterborne Commerce of the United States:
Calendar Year 1978.

(Table 1 continues on next page.)

Table 1 (Cont.)
1978 PETROCHEMICAL MOVEMENTS ON THE MISSISSIPPI RIVER
FROM MOUTH OF PASSES TO THE PORT OF BATON ROUGE
BY COMMODITY AND SHORT TON

COMMODITY	RECEIPTS	COASTWISE SHIPMENTS	THROUGH UPBOUND	DOWNBOUND
Crude petroleum	5,789,674	338,824	5,781,014	91,212
Basic chemicals	27,630	196,689	11,713	184,722
Gasoline	203,655	4,167,304	181,016	3,677,697
Distillate fuel oil	253,987	2,241,415	175,163	2,067,940
Residual fuel oil	950,923	3,070,116	587,437	2,840,178
Lubricating oil	41,697	258,184	22,813	248,402
Coke, petroleum coke				
GRAND TOTAL:	7,267,566	10,272,532	6,759,156	9,110,143

SOURCE: Department of Army Corps of Engineers, "Part 2: Waterways and Harbors-Gulf Coast Mississippi River System and Antilles," Waterborne Commerce of the United States: Calendar Year 1978.

(Table 1 continues on next page.)

Table 1 (Cont.)
1978 PETROCHEMICAL MOVEMENTS ON THE MISSISSIPPI RIVER
FROM MOUTH OF PASSES TO THE PORT OF BATON ROUGE
BY COMMODITY AND SHORT TON

	INTERNAL[a]		
COMMODITY	RECEIPTS	SHIPMENTS	LOCAL
Crude petroleum	6,206,358	8,406,554	1,362,419
Basic chemicals	1,015,372	3,635,908	1,309,596
Gasoline	1,580,063	5,364,922	431,697
Distillate fuel oil	934,836	3,559,556	1,846,306
Residual fuel oil	5,723,988	6,853,199	3,978,830
Lubricating oil	297,881	1,472, 588	23,278
Coke, petroleum coke	535,765	323,147	79,180
GRAND TOTAL:	16,294,263	29,615,874	8,231,306

[a] Internal traffic data was not available for movements completely
within the stretch between mile 127 (Norco, LA) and mile 168
(Burnside, LA).

SOURCE: Department of Army Corps of Engineers, "Part 2: Waterways
and Harbors-Gulf Coast Mississippi River System and
Antilles," Waterborne Commerce of the United States:
Calendar Year 1978.

Of the total tanker traffic recorded on the lower section of the Mississippi River in 1978 (3850 ship movements), approximately 45% (1728) consisted of vessels with a draft of 30 feet or more. Approximately 1047 shipments (27.2%) were made in tankers with a draft of 35 feet or greater, and 260 of those were in tankers with a draft of 40 feet--the maximum draft capable of navigating the existing channel (see Table 2).

Table 2

TANKER MOVEMENTS WITH 30 TO 40-FOOT DRAFTS
BETWEEN MOUTH OF PASSES AND THE
PORT OF BATON ROUGE, 1978

Drafts	Inbound	% of Total	Outbound	% of Total	Total	% of Total
30-40	859	44.6	869	45.2	1,728	44.9
35-40	608	31.5	439	22.8	1,047	27.2
40	178	9.2	82	4.2	206	6.8
Total Trips	1,927	100.0	1,932	100.0	3,850	100.0

SOURCE: Department of the Army Corps of Engineers, "Part 2: Waterways and Harbors--Gulf Coast Mississippi River System and Antilles," Waterborne Commerce of the United States: Calendar Year 1978.

The transportation of crude or products in large vessels is more cost effective than comparable movements in smaller vessels. The current World Scale rates (April 27, 1981) from selected origins to a destination between the ports of New Orleans and Baton Rouge are illustrative of this point. Assuming the chartering of a tanker for an average grade of crude, the scales in Table 3 are applicable:

Table 3
WORLD SCALE ORIGIN

Tanker Size K DWT	Persian Gulf	West Coast Africa	North West Europe	Mediter- ranean	Venezuela
80	80	80	70	75	85
150	47.5	47.5	40	42.5	52.5
% Difference	40.6	40.6	42.9	43.3	32.4

. It should be noted, however, that when the larger tankers (greater than 80K DWT) are used to transport crude into the area, they must be lightered to meet the navigation constraints of the Mississippi River. The lightering process (discussed below) is accomplished by off-loading cargo either to barges or smaller ships (Exhibit 3) and is both a time-consuming and expensive operation which costs up to $1.00 per barrel of crude transferred.

In 1980, 1294 vessels carrying a combined total of 490,977,641 barrels of crude oil entered the Mississippi River to deliver to facilities between the Mouth of Passes and the Port of Baton Rouge (see Table 4 for countries of origin of imported crude). Fifty percent of these ships received their oil from larger tankers in the Gulf. Of those vessels that entered the Mississippi River, 3% were forced to lighter into barges to reach their port of destination. It should be noted, however, that the number of vessels required to off-load into barges is expected to be significantly greater. One contractor alone reported the lightering of 13 vessels (approximately 1,300,000 barrels of crude) in the period from January 1, to March 31, 1981. In 1980 an additional 2% of the ships coming into the river lightered into some type of smaller vessel other than barges.

Table 4

COUNTRY OF ORIGIN OF IMPORTED CRUDE PETROLEUM
RECEIVED ON THE MISSISSIPPI RIVER FROM THE
MOUTH OF PASSES TO THE PORT OF BATON ROUGE

Country	Barrels
Saudi Arabia	19,148,128
Bahamas	9,819,553
Algeria	54,498,518
Angola	14,729,329
Libya	71,191,758
Mexico	50,750,697
Nigeria	155,639,912
Norway	13,747,374
Trinidad	28,967,681
United Kingdom (North Sea)	49,588,742
Venezuela	23,075,949
TOTAL	490,977,641

Source: U.S. Customs Service, New Orleans, Louisiana

The average payload of the tankers carrying crude into the Mississippi River in 1980 was approximately 379,500 barrels (490,977,641 barrels imported in 1294 vessels). If the 50% of these shipments that are lightered from tankers in the Gulf could have been transported to the refineries in larger ships, the traffic on the river could have been reduced by 424 vessels. That is, 647 received oil offshore from larger ships; if that volume of crude, 245,536,500 barrels (647 x 379,500), could have been transported in vessels with a capacity of 1,100,000 barrels (average 150K DWT capacity), this movement would have been made in only 223 tankers.

The cost of shipping oil in small tankers relative to importing it in larger tankers has been researched. If the crude petroleum and No. 6 fuel oil imported in 1980 had been shipped in 80K DWT tankers from their ports of origin to their destinations on the lower Mississippi River, the cost of transportation would have been approximately $536,300,000 and $14,600,000, respectively. On the other hand, if these commodities had been shipped in 150K DWT tankers, the estimated total cost of transportation would have been $322,600,000. This represents a savings of approximately $228,300,000. The details of these calculations are discussed later.

The cost of transporting these commodities via 150K DWT tankers and lightering at the Mouth of Passes was also researched. A cost computation was made to reflect the long haul transportation costs of a large tanker (VLCC) and the required lightering cost to allow the VLCC to navigate the river channel to its point of destination. The final cost estimate of $555,251,100, which is also analyzed later, approximates that computed for the transportation of crude and fuel oil in 80K DWT tankers from point of origin to point of destination.

The scenarios described above are presented as illustrations of the potential of deep draft vessels. They are based upon average payloads and a number of other constructed assumptions and are not intended to be definitive statements about crude movements on the lower Mississippi River.

Refineries in the project area anticipate a continuing demand for crude oil, and this expectation is supported by the projections presented in documents prepared for the formation of the Third National Energy Plan. It is projected that imports will continue either at or near present levels through the year 1990. Present levels are 7.9 million barrels per day, of which approximately 1.975 barrels (25%) enter the U.S. through the Mississippi River).

Preliminary National Energy Plan III, base case projections, prepared by DOE in November, 1980, assumed existing legislation and continuation of the previous Administration's energy policies as of that date. In these base case estimates, U.S. oil imports are likely to range from 7 to 9 million barrels per day in 1985, and 4 to 8 million barrels per day in 1990. Currently about 7.9 million barrels per day are imported. To deal with this amount of oil in an efficient manner, we shall have to address certain bottlenecks, one of the most troublesome of which is "lightering."

LIGHTERING

Lightering is the process used to raise the draft of a vessel in which part of the cargo is off-loaded onto another vessel. This procedure is undertaken on a regular basis by crude-oil tankers entering the Mississippi River. Two methods of lightering are employed by the oil industry in the study area. The first method involves off-loading a portion of the cargo onto a contract barge or smaller vessel and then transporting the cargos in both vessels to the refinery. In the other method, a larger vessel unloads its cargo offshore onto smaller vessels which can navigate the river's restricted channel.

Of the two processes, it is difficult to determine which is more costly. The use of multiple vessels, however, can assuredly be categorized as an anticipated expense. In contrast, the use of the barge method is often unplanned, and thus its cost is unexpected. The latter method must be utilized primarily when a vessel's draft exceeds that of the channel on a given day, although it is also used on a planned basis. The former method is employed to lessen the cost of long distance transportation by use of very large ships. Smaller vessels then shuttle cargo to refineries from the Gulf. The cost of both types of lightering operations contributes directly to the price of refining oil and places rather strict restraints on the operations of refining facilities. A deeper channel would affect these lightering procedures by allowing larger, more cost-effective vessels to enter the river without the necessity of planned lightering.

The estimated cost of lightering onto barges, a contract service available through independent contractors in the area, approaches and sometimes exceeds $1.00 per barrel off-loaded. Average lightering jobs transfer between 100,000 to 200,000 barrels. Generally, this cost is reflected as an additional $.20--$.35 per barrel on the original shipment. Based on an average shipment of 600,000 barrels, this cost would be significant. Demurrage charges, ranging from 20,000-35,000 per day, must often also be paid for, delaying the normal operation of the vessel's contract in order to off-load cargo.

Estimations of ship-to-ship lightering costs cannot be readily ascertained. Prices vary according to contract and charter agreements. It is not unreasonable, however, to envision the costs as easily approaching those of the barge lightering operations.

POTENTIAL TRANSPORTATION COSTS OF IMPORTING CRUDE PETROLEUM AND NO. 6 FUEL OIL IN 150K DWT AND 80K DWT TANKERS

In order to compare the costs of shipping crude petroleum and selected products by 150K DWT and 80K DWT tankers from the origins of the shipments to their destination on the lower Mississippi River, these transportation costs were estimated using the current World Scale Freight and World Scale Vessel Chartering Rates in the following manner:

1. Current rates were obtained for the applicable movements. The World Scale Freight Rates are set biannually and represent base costs per ton to ship a particular commodity between two specified ports. Vessel chartering rates fluctuate according to the demand for, and supply of, each type of vessel within an area. Using the World Scale Freight Rate as the base unit, a vessel chartering rate represents the fraction of this unit which must be paid to transport a commodity in a vessel of specified dimensions. The applicable World Scale Rates for shipping an average grade of crude oil from a representative port in each of the seven regions which export crude to the study area are displayed in Table 5. (Each of the countries listed in Table 4 as an exporter of crude to the lower Mississippi River, with the exception of the Bahamas, is included within one of these seven regions.) No petroleum is produced in the Bahamas, but crude is stored there for transshipment. Thus, for the purposes of our calculations, crude imported from the Bahamas was considered to have originated in one of the exporting regions on the eastern side of the Atlantic Ocean in approximately the same proportions as crude that is shipped directly from those regions.

2. The number of barrels of crude imported from each of the areas of production was converted to long tons using a ratio of 8.4 barrels of crude to one long-ton.

3. The tonnages coming from each region were multiplied by the World Scale Freight Rate to get the base cost of shipment.

4. In order to obtain the cost of transporting the shipments in 150K DWT tankers and 80K DWT tankers, the resulting values were multiplied by current World Scale Vessel Chartering Rates.

5. The difference between the calculated cost of shipment of crude oil in the two classes of tankers was listed in Table 5 for imports from each region.

6. Import figures of refined petroleum products were studied
    to determine whether transporting those commodities in larger
    tankers might reduce transportation costs. As can be seen from
    Table 6, the only product which was imported in sufficient
    quantities in 1980 to realize a substantial savings from
    increasing the volumes of transporting vessels was No. 6 fuel
    oil.
7. The costs of shipping this No. 6 fuel oil, which is imported
    primarily from Algeria, were calculated for 150K DWT and 80K
    DWT vessels in the same manner as they had been for crude
    petroleum. For these calculations, it was assumed that the
    World Scale Rates for transporting fuel oil from the
    Mediterranean are not substantially different from those for
    shipping petroleum. No. 6 fuel oil is heavier than crude
    petroleum, however, and thus a different factor was used to
    convert the number of barrels imported to long tons.
8. The total costs of transporting crude and No. 6 fuel oil in
    each of the two classes of tankers were found and are listed in
    Table 5 along with the difference between the two total cost
    figures.

POTENTIAL TRANSPORTATION COSTS OF
IMPORTING CRUDE PETROLEUM
AND NO. 6 FUEL OIL BY 150K DWT TANKERS AND LIGHTERING
TO ALLOW NAVIGATION OF THE MISSISSIPPI RIVER

As a second point of comparison, the transportation cost was
computed for crude shipped from the port of origin in 150K DWT
tankers which were lightered to the point that the ship could
navigate the existing channel at the mouth of the river. It was
assumed that a fully loaded 80K DWT vessel carrying an equivalent
amount of crude would also be capable of passing through the
channel. Thus the difference in the amount of payload between the
two vessels was assumed to be the volume lightered. Under the
assumptions of this scenario, approximately 46.4% of the crude
transported in the 150K DWT tankers would be lightered at the
current estimated cost of $1.00 per barrel. This figure for the
volume of crude lightered is comparable to the amount reported by
the U.S. Customs in 1980. Continuing the calculations using this
volume figure results in a relative cost estimate that is consistent
with current experience in ship chartering.

It is very interesting to note that the total cost of transporting crude in this scenario is $555,251,100, an amount that is
remarkably close to the calculated cost of shipping the crude in 80K

Table 5

POTENTIAL TRANSPORTATION COSTS OF IMPORTING CRUDE
PETROLEUM AND NUMBER 6 FUEL OIL IN 150K AND 80K DWT TANKERS

ORIGIN	WORLD SCALE FREIGHT RATE (DOLLARS)	WORLD SCALE VESSEL CHARTERING RATES	
		150K DWT TANKER	50K DWT TANKER
CRUDE PETROLEUM[1]			
PERSIAN GULF	21.58	.475	.80
WEST COAST AFRICA			
NIGERIA	14.37	.475	.80
ANGOLA	15.26	.475	.80
N.W. EUROPE	11.36	.400	.70
MEDITERRANEAN	13.87	.425	.75
VENEZUELA	4.74	.525	.85
MEXICO	2.56	.420	.60
TOTAL CRUDE PETROLEUM			
NUMBER 6 FUEL OIL			
MEDITERRANEAN	13.87	.425	.75

[1] Represents quantities imported in 1980.

(Table 5 continues on next page.)

Table 5 (Cont.)
POTENTIAL TRANSPORTATION COSTS OF IMPORTING CRUDE
PETROLEUM AND NUMBER 6 FUEL OIL IN 150K AND 80K DWT TANKERS

ORIGIN	QUANTITY BARRELS	LONG TONS	COST OF[2] SHIPMENT (DOLLARS)
CRUDE PETROLEUM[1]			
PERSIAN GULF	19,648,900	2,339,200	50,470,000
WEST COAST AFRICA			
NIGERIA	159,675,700	19,009,000	273,159,600
ANGOLA	15,112,300	1,799,100	27,454,000
N.W. EUROPE	64,976,000	7,735,200	87,872,300
MEDITERRANEAN	128,950,400	15,351,200	212,921,600
VENEZUELA	52,043,600	6,195,700	29,365,500
MEXICO	50,470,700	6,020,300	15,412,000
TOTAL CRUDE PETROLEUM	490,977,600	58,449,700	
NUMBER 6 FUEL OIL			
MEDITERRANEAN	10,509,800	1,401,300	19,436,100
TOTAL	510,487,400	59,851,000	

[1] Represents quantities imported in 1980.
[2] At prevailing World Scale Freight Rates.

(Table 5 continues on next page.)

Table 5 (Cont.)
POTENTIAL TRANSPORTATION COSTS OF IMPORTING CRUDE
PETROLEUM AND NUMBER 6 FUEL OIL IN 150K AND 80K DWT TANKERS

ORIGIN	COST OF SHIPMENT[2] BY VESSEL (DOLLARS)		COST DIFFER-ENCE BETWEEN METHODS OF SHIPMENT (DOLLARS)
	150K DWT TANKER	80K DWT TANKER	
CRUDE PETROLEUM[1]			
PERSIAN GULF	23,977,500	40,383,200	16,405,700
WEST COAST AFRICA			
NIGERIA	129,750,800	218,527,760	88,776,900
ANGOLA	13,040,700	21,963,200	8,922,500
N.W. EUROPE	35,148,900	61,510,600	26,361,700
MEDITERRANEAN	90,491,700	159,691,200	69,199,500
VENEZUELA	15,417,900	24,962,400	9,544,500
MEXICO	6,473,100	9,247,200	2,774,100
TOTAL CRUDE PETROLEUM	314,300,600	536,285,500	221,984,900
NUMBER 6 FUEL OIL			
MEDITERRANEAN	8,260,300	14,577,100	6,316,700
TOTAL	325,509,900	550,862,600	228,301,600

[1] Represents quantities imported in 1980.
[2] At prevailing World Scale Freight Rates multiplied by the
World Scale Vessel Chartering Rates.

Table 6
IMPORTS OF REFINED PETROLEUM PRODUCTS
TO THE MISSISSIPPI RIVER, MOUTH OF
PASSES THROUGH THE PORT OF BATON ROUGE

CARGO	BARRELS	POUNDS	PRIMARY COUNTRY OF ORIGIN
Creosote	678,129		Trinidad
Catfeed	1,958,000		Nigeria, Libya
Zylene	112,892		Trinidad
Toulene	897,411		Libya
Premium Gas	47,266		United Kingdom
No. 6 Fuel Oil	10,509,798		Algeria
Naptha		108,497,011	Argentina
Benzene		12,100,089	Israel
Butidiene		41,821,749	

Source: U.S. Customs Service, New Orleans, Louisiana

DWT tankers. Given the elasticity of demand in ship chartering, it is reasonable to assume that the cost of transporting crude by these two methods would be similar. If these costs were not similar, the cost differential would result in the exclusive or predominate use of one mode. The transportation cost for this scenario was computed as follows:

1. It was assumed that 490,977,600 barrels of crude and 10,509,800 barrels of No. 6 fuel oil would be transported in 150K DWT tankers at current rates from those ports of origin recorded in 1980. As was demonstrated earlier, the total cost of this movement was found to be $322,560,900.

2. It was estimated that 46.4% (232,690,200 barrels) of the crude and No. 6 fuel oil would be lightered at $1.00 per barrel.

3. Therefore, the cost of lightering and the cost of shipment in 150K DWT tankers were summed, and the result was established as the total cost of transporting the crude of fuel oil from its points of origin to its destinations on the Mississippi River.

NOTES

CHAPTER III
A REMEDIABLE IMPEDIMENT TO EXPORT SALES
OF U.S. HIGHER SULFUR COALS
THOMAS TRUMPY

The technical information in this article was derived from information made available by Combustion Engineering-Raymond. The author wishes to thank Messrs. Larry Morton and Bill Converse of Combustion Engineering Europe, Paris; Mr. Zachariah Allen of F.R. Schwab & Associates, New York; and Mr. Richard Sommer of New Canaan, Connecticut, for their contributions to this note.

CHAPTER III
THE POTENTIAL WORLD DEMAND
FOR HIGH SULFUR COAL
RAY LONG

The views expressed in this paper are those of the author; they reflect the economic assessment work going on in IEA Coal Research, but are not necessarily the views of the organization or of supporting countries.

CHAPTER III
MARKETING MECHANISMS FOR THE
EXPORT OF HIGH SULFUR COAL
ARTHUR F. NICHOLSON

NOTES

[1] "The State of the Environment in OECD Member Countries" (Paris: OECD, 1979).

[2] "Clean Fuel Supply" (Paris: OECD, 1979).

[3] Barnes, R. A. "The Long Range Transport of Air Pollution. A Review of European Experience." Journal, Air Pollution Control Association 29(1979):1219-1235.

[4] Rubin, E. S. "Air Pollution Constraints on Increased Coal Use by Industry. An International Perspective." Journal, Air Pollution Control Association 31(1981):349-360.

CHAPTER III
ILLINOIS COAL EXPORT POTENTIAL
AND INTERNATIONAL PRICE COMPETITIVENESS
C. KENT GARNER

1. U.S. Department of Energy Interagency Coal Export Task Force. Interim Report of the Interagency Coal Export Task Force. (Draft for public comment): 3-1.

2. Ibid., p. 3-16, 18.

3. Coal Week. 7(1-20) and 6 (44-51).

4. Coal Week International. 2(16):5 and 2(15):4.

5. Electric Power Research Institute. "Coal Industry Problems." Coal Week 7(17):10.

6. If Illinois coal were priced on an energy equivalent basis with South African coal:

    ($2.46/MM Btu) x (11,700 Btu/lb) x (2,000 lb/s.t.) x (1/1,000,000)

FAS European Piers (U.S. $ per short ton)	57.56
Less:	
Ocean freight	(13.00)
Port, loading/unloading charges	(4.50)
Railroad/barge freight	(12.50)
F.O.B. Mine Price	$ 27.56

7. Daniel Klein, ICF, Washington, D.C., statement at Coal Week/Energy Bureau Conference. Coal Week 6(42):7.

8. Productivity Report--Fourth Quarter 1980 and Preliminary Year 1980 for All Producers by State, County and Type Mining. (Washington, D.C.: Pasha Publications, March 25, 1981).

NOTES ON CONTRIBUTORS

Zachariah Allen is a Senior Vice President with F. R. Schwab and
   Associates, Inc., where he heads the Strategic Services Group.
   A Harvard graduate with an MBA in finance, he conducts
   corporate planning, marketing, and strategic development
   studies for foreign and domestic clients in the energy
   industries.

Carl E. Bagge has, since 1971, been President of the National Coal
   Association, the industry trade association representing major
   commercial bituminous coal producers, coal sales companies,
   mining equipment manufacturers coal-carrying railroads and
   barge lines, and resource developers. The author of more than
   fifty articles and holder of a J.D. degree from Northwestern,
   Mr. Bagge has also served as a member and Vice Chairman of the
   Federal Power Commission and as General Attorney for the
   Atchison, Topeka and Santa Fe Railroad.

Peter Borre, Acting Assistant Secretary for International Affairs at
   the Department of Energy, has recently been in charge of energy
   negotiations with Venezuela, Israel, Italy, and Nigeria. He
   has held energy policy-planning positions in the federal
   government since 1974 and is a frequent witness before
   congressional committees. Mr Borre holds an MBA from the
   Harvard Business School.

Harry J. Bruce is Senior Vice-President of Marketing for the
   Illinois Central Gulf Railroad, a position he has held since
   1975. The author of more than fifty articles and two books,
   Mr. Bruce has served in marketing, transportation or
   distribution positions for U.S. Steel, Spector Freight System,
   the Joseph Schlitz Brewing Company, as well as the Western
   Pacific Railroad. He holds an M.S. from the University of
   Tennessee and has completed the Advanced Management Program at
   Harvard.

K. S. Chang is Minister of Economic Affairs, The Republic of China.

J. Alan Cope is Assistant Vice-President of Marketing for
   Consolidation Coal Company.

Michael Crow is the former Assistant Director for Program Planning and Analysis at Southern Illinois University's Coal Research Center. Involved in coal research since 1974, Mr. Crow has served in research, planning and management positions in several organizations. Currently Mr. Crow is serving as a Research Fellow in the Science and Technology Policy Program at the Maxwell School, Syracuse University.

Gerald D. Cunningham is Director of Marketing for International Matex Coal Company's export facility near New Orleans. Before entering the terminal industry, Mr. Cunningham's background included various financial and transportation positions, including six years as Regional Office Manager for Illinois Central Gulf Railroad's New Orleans office.

Willem G. Daniels is Director of the North American Delegation of the Association Technique de l'Importation Charbonnievve (A.T.I.C.), New York. He has served in a variety of other shipping positions, most recently as Manager/Director at A.T.I.C.'s Paris office, where he was responsible for coal purchasing from various countries, including all contracting with U.S. coal suppliers. Mr. Daniels studied for two years at the Dutch Institute for Foreign Service and has a Baccalaureat es Sciences from the Academie de Paris.

Donald V. Earnshaw is Deputy Assistant for Export Development in the U.S. Department of Commerce.

George Ecklund is Vice-President of Zinder-Neris, Inc. Dr. Ecklund's consulting expertise includes domestic and international marketing studies for both steam and metallurgical coal on the micro and regional basis.

Wayne T. Ewing is President of the Illinois Division of the Peabody Coal Company. Beginning as a laborer in 1963, Mr. Ewing progressed through various positions at Peabody (including three vice-presidencies) and was most recently President of the Company's Indiana Division. Mr. Ewing has an M.A. in Education from Western Kentucky University.

Jean M. Faucounau is Director of the Permanent Office of Charbonnages de France in Washington. A graduate of the Ecole Polytechnique, Mr. Faucounau has also been Chief of the Chemical Plant of Carling, General Director of the Societe Chimique de l'Adour, and General Manager of the Societe des Engrais de l'Ile de France.

John P. Ferriter is Director of the Office of Energy Consuming Countries in the Economic and Business Bureau of the Department of State. His duties include U.S. relations with the International Energy Agency. A career Foreign Service officer since 1964, Mr. Ferriter recently served in the office of International Commodities at the Department of State. He holds an MPA from Harvard University.

C. Kent Garner is Vice-President and Head of the U. S. Section of Continental Illinois National Bank's Mining Division, a firm he has been associated with since 1973. He has worked with midwestern and Appalachian coal mining companies, managed Continental's Corporate Development Division and served in the bank's Sydney, Australia, office as its representative for energy and mineral resources. Mr. Garner has an MBA from the University of California, Berkeley.

Herbert Haar has been the Assistant Executive Director for Planning and Engineering for the port of New Orleans since 1971. He also serves on the National Academy of Sciences Committee on the State Role in Waterborne Transportation and is Chairman of the American Association of Port Authorities Ad Hoc Dredging Task Force. Colonel Haar has an M.S. degree from the University of Illinois.

Dr. Michael O. Holowaty is presently Senior Advisor, Research Department, at the Inland Steel Company of East Chicago. He has held various professional and managerial positions in that Department and recently received the 1980 Joseph Becker Award for his work on the development of Illinois coal for metallurgical applications. Dr. Holowaty received his Ph.D. in inorganic chemistry from the University of Breslau, Germany, in 1944.

Robert Jackson, Manager of the Commercial Division of the Conoco Coal Development Company, is a member of the International Symposium on Alcohol Fuel Technology. He has been involved with the development of alcohol fuels during the past nine years and prior to that worked in the R & D department of Shell Research Ltd. Mr. Jackson is a graduate of the University of Liverpool.

Akiro Kinoshita is presently Senior Economist for the Department of Fuel and the Department of General Planning in the Electric Power Development Company of Tokyo. He has also been a member of the Study Committee on Imported Coal for the Institute of

Energy Economics and an Associate and Coordinator for the
M.I.T. World Coal Study. He is a 1957 graduate of Keio
University.

Dr. Ulf Lantzke, since 1974, has been the Executive Director of the
International Energy Agency, Paris, which is composed of 21
industrial nations working together to solve their energy
problems. Previous to this he was Head of the Energy Department and responsible for energy policy, iron and steel, and raw
materials in the Federal Ministry for Economic Affairs in Bonn.
He received a J. D. degree in law from the University of
Munster in 1952.

Dr. Bernard S. Lee, President of the Institute of Gas Technology
since 1978, is responsible for all IGT operations. Previous to
his present position, he served as IGT's Executive
Vice-President and also as Vice-President and Head of the
Process Research Division. He holds a doctorate in chemical
engineering from the Polytechnic Institute of New York.

Bong Suh Lee is Assistant Minister in the Ministry of Energy and
Resources in Seoul, Republic of Korea.

Roger W. A. LeGassie is currently DOE's Acting Assistant Secretary
for Fossil Energy, a position he assumed January, 1981. He is
also Acting Director of the DOE office of Policy Planning and
Analysis. Mr. LeGassie began his federal service in 1952, and
has since received several special achievement awards, among
them the AEC's Distinguished Service Award. He holds a B.A.
degree in mathematics and chemistry from Columbia University.

Ray Long, Energy Economist since 1979 for International Energy
Agency (IEA) Coal Research, London, is an authority on global
energy flows and shipping, and has recently developed a compu-
terized energy model to help forecast energy supplies. He has
previously served as a consultant in shipping and energy
economics for financial institutions. Mr. Long holds a B.A. in
economics and geography from Emmanuel College, Cambridge.

Ramesh Malhotra is Assistant Vice-President for Marketing for
Freeman United Coal Mining Company where he oversees and
directs the company's coal marketing programs. Prior to
assuming this position in 1977, Mr. Malhotra was Director of
Market Planning at Freeman, and before that he was Principal
Mineral Economist for the Illinois State Geological Survey.

Mr. Malhotra has an M.S. in business administration from Michigan Technological University.

William W. Mason is President of Island Creek Coal Sales Company and serves also as President of the Coal Exporters Association. Prior to assuming his firm's presidency in 1978, Mr. Mason served in a variety of positions with Island Creek. He received an engineering degree from Marshall University in 1947.

Arthur F. Nicholson, Commissioner of the Bureau of Energy Production and Utilization in the Kentucky Department of Energy, is responsible for the development of policies and programs to enhance the production and utilization of coal and other forms of energy. A retired Air Force officer with experience in the management of large transportation units, he is the author of numerous marketing and transportation reports and has previously served as a Professor of Aerospace Studies at the University of Kentucky and New York University.

The Honorable Charles H. Percy was elected to the U.S. Senate in 1966 and re-elected in 1972 and 1978. He is Chairman of both the Foreign Relations Committee and the Governmental Affairs Subcommittee on Energy, Nuclear Proliferation, and Government Processes. The Senator's major legislative initiatives and accomplishments have focused on government reform, the economy, the elderly, and energy. He has been a strong advocate of legislation to forge a national energy policy. Senator Percy is a graduate of the University of Chicago with a degree in economics.

Michael K. Reilly is President of Ziegler Coal Company.

Dr. Lyle V. A. Sendlein, Professor of Geology, was Director of the Coal Research Center at Southern Illinois University at Carbondale, from 1977-1982. He is presently Director of the Institute for Mining and Minerals Research at the University of Kentucky. Dr. Sendlein holds a Ph.D. in geology from Iowa State University.

Jose Sierra-Lopez is President of CARBOEX, Spain.

Paul Simon, U.S. Congressman from the 24th District of Illinois, began his professional career as a journalist before serving in the Illinois House and Senate. He was Lieutenant Governor of the State of Illinois from 1969-1973 and was elected to the

U.S. House of Representatives in 1974 and subsequently re-elected. Congressman Simon is the author of five books and holds seven honorary doctorates.

Thomas Trumpy is European Director of Project Assistant International.

J. Harley Williams is Vice-President of Marketing for the Old Ben Coal Company, a position he has held since 1973. His responsibilities include the sale, transportation, distribution, and market planning for approximately two million tons of coal annually. Prior to his present position, Mr. Williams was Senior Attorney for Sohio where he was responsible for legal affairs in the areas of coal and other natural resources. He holds a J.D. degree from the University of Colorado.